# 9 Crystals

**Growth, Properties, and Applications**

---

Managing Editor: H. C. Freyhardt

Editors: T. Arizumi, W. Bardsley, H. Bethge
A. A. Chernov, H. C. Freyhardt, J. Grabmaier
S. Haussühl, R. Hoppe, R. Kern, R. A. Laudise
R. Nitsche, A. Rabenau, W. B. White
A. F. Witt, F. W. Young, Jr.

# Modern Theory of Crystal Growth I

Editors: A. A. Chernov and H. Müller-Krumbhaar

With Contributions by
P. Bak A. Bonissent J. van der Eerden
W. Haubenreisser H. Pfeiffer V. V. Voronkov

Springer-Verlag
Berlin Heidelberg New York 1983

**Managing Editor**

Prof. Dr. H. C. Freyhardt, Kristall-Labor der Physikalischen Institute, Lotzestr. 16–18, D-3400 Göttingen
and Institut für Metallphysik der Universität Göttingen, Hospitalstr. 12, D-3400 Göttingen

**Editorial Board**

Prof. *T. Arizumi,* Department of Electronics, Nagoya University, Furo-cho Chikusa-Ku, Nagoya 464, Japan
Dr. *W. Bardsley,* Royal Radar Establishment, Great Malvern, England
Prof. *H. Bethge,* Institut für Festkörperphysik und Elektronenmikroskopie, Weinberg, 4010 Halle/Saale, DDR
Prof. *A. A. Chernov,* Institute of Cristallography, Academy of Sciences, Leninsky Prospekt 59, Moscow B – 11 73 33, USSR
Dr. *J. Grabmaier,* Siemens AG, Forschungslaboratorien, Postfach 80 17 09, 8000 München 83, Germany
Prof. *S. Haussühl,* Institut für Kristallographie der Universität Köln, Zülpicherstr. 49, 5000 Köln, Germany
Prof. *R. Hoppe,* Institut für Anorganische und Analytische Chemie der Justus-Liebig-Universität, Heinrich-Buff-Ring 58, 6300 Gießen, Germany
Prof. *R. Kern,* Université Aix-Marseille III, Faculté des Sciences de St. Jérome, 13397 Marseille Cedex 4, France
Dr. *R. A. Laudise,* Bell Laboratories, Murray Hill, NJ 07974, U.S.A.
Prof. *R. Nitsche,* Kristallographisches Institut der Universität Freiburg, Hebelstraße 25, 7800 Freiburg, Germany
Prof. *A. Rabenau,* Max-Planck-Institut für Festkörperforschung, Heisenbergstr. 1, 7000 Stuttgart 80, Germany
Prof. *W. B. White,* Materials Research Laboratory, The Pennsylvania State University, University Park, PA 16802, U.S.A.
Prof. *A. F. Witt,* Massachusetts Institute of Technology, Cambridge, MA 02139, U.S.A.
Dr. *F. W. Young,* Jr., Solid State Division, Oak Ridge National Laboratory, P.O. BOX X, Oak Ridge, TN 37830. U.S.A.

**Guest Editor**

Prof. Dr. H. *Müller-Krumbhaar*, Institut für Festkörperforschung, KFA Jülich, Postfach 1913, 5170 Jülich, Germany

ISBN 3-540-12161-7 Springer-Verlag Berlin Heidelberg New York
ISBN 0-387-12161-7 Springer-Verlag New York Heidelberg Berlin

Library of Congress Cataloging in Publication Data. Main entry under title: Modern theory of crystal growth I. (Crystals – growth, properties, and applications; 9) 1. Crystals – Growth – Addresses, essays, lectures. I. Chernov, A. A. II. Müller-Krumbhaar, H. III. Bak, P. (Per), 1947– . IV. Series. QD921.C79 1978 vol. 9 548 s [548'.5] 83-372 ISBN 0-387-12161-7 (U.S.)

This work is subject to copyright. All rights are reserved, whether the whole or part of materials is concerned, specifically those of translation, reprinting, re-use of illustrations, broadcasting, reproduction by photocopying machine or similar means, and storage in data banks. Under § 54 of the German Copyright Law where copies are made for other than private use, a fee is payable to "Verwertungsgesellschaft Wort", Munich.

© by Springer-Verlag Berlin Heidelberg 1983. Printed in Germany

The use of registered names, trademarks, etc. in this publication does not imply, even in the absence of a specific statement, that such names are exempt from the relevant protective laws and regulations and therefore free for general use.

Typesetting and printing: Schwetzinger Verlagsdruckerei. Bookbinding: K. Triltsch, Würzburg.
2152/3140-5 4 3 2 1 0

# Preface

Our understanding of the basic processes of crystal growth has meanwhile reached the level of maturity at least in the phenomenological concepts. This concerns for example the growth of pure crystals from a low-density nutrient phase like vapor or dilute solution with various aspects of pattern formation like spiral and layer growth, facetting and roughening, and the stability of smooth macroscopic shapes, as well as basic mechanisms of impurity incorporation in melt growth of (in this sense) simple materials like silicon or organic model substances. In parallel the experimental techniques to quantitatively analyze the various growth mechanisms have also reached a high level of reproducibility and precision, giving reliable tests on theoretical predictions. These basic concepts and applications to experiments have been recently reviewed by one of us (A.A.C.) in "Modern Crystallography III. Crystal Growth" (Springer Series on Solid State Sciences, 1983).

It has to be emphasized, however, that for practical applications we are still unable to quantitatively calculate many important parameters like kinetic coefficients from first principles. For mixed systems such as complex oxides, solutions and systems with chemical reactions, our degree of understanding is even lower. As a few examples for present achievements we note that experiments with vapour and molecular beam condensation of alkali halides confirmed the qualitatively predicted mechanisms of screw dislocations and two-dimensional nucleation for layer-growth. The same holds for precisely controllable experiments of electro-crystallization of silver crystals or conventional growth of ADP crystals in aqueous solution under simultaneous X-ray topographic control. Here the functional relations of growth rate depending upon external parameters are confirmed, while absolute values still have to be fitted between theory and experiments.

In at least three major aspects, furthermore, the achievements of crystal growth theory so far appear to be still unsatisfactory, leading us to the conclusion, that this field is only at the beginning of important developments.

The first point is, that even the present "microscopic" models treat the elements of a crystal (atoms or molecules) as "brickstones", which are accumulated to yield a regular solid lattice. This concept completely ignores the subtle electronic rearrangements taking place under the incorporation of an additional particle into a crystal surface, containing the interactions only as short-ranged effective two-body forces. This is clearly not an extremely good concept for example for metals, where a substantial fraction of the electrons is not localized. Comparable in difficulty albeit technically distinct are questions of growing ionic crystals or molecular crystals.

The second point is, that there exists no generally accepted theory of melting and freezing so far. The basic reason here is the difficulty of properly handling the translational and orientational degrees of freedom of the liquid and specifically the spontaneous breaking of these symmetries in the freezing transition. Atomistic models for the solid-liquid interface, therefore, are still rudimentary, but progress is in sight and the first two chapters of this volume are dealing with such atomistic models.

As a consequence of this lack of understanding atomistic systems, our knowledge on the formation of defects and dislocations is limited. Even the problem of discriminating between regular and defected interface sites as nucleation sources is not solved. We feel, therefore, that the importance of defect and dislocation formation for crystal growth will make this a central issue for further investigations. The use of large-scale computer simulations can be expected to become increasingly important in this field.

The third point, finally, concerns the simultaneous consideration of interface structure and transport processes like diffusion and convection. The interface structure and composition is largely determined by those macroscopic transport mechanisms, taking place at completely different length scales. In addition, convection in a melt just by itself is already a problem which is hard to handle analytically.

One important application concerns the formation of striations in Czochralski growth. There it is still not clear, whether the oszillating appearance of impurities in the crystal simply reflects the time scales of the convection patterns or envolves a complicated interplay between kinetic incorporation, diffusion and convection.

While a satisfactory treatment of the first of these three complexes can only be expected in the more distant future, the latter two points are being attacked with increasing effort in the present. Our aim for this and the following volume is, therefore, to present a collection of articles that show todays pathways into these complicated and fascinating problems of ordered solid lattices growing from homogeneous nutrient phases.

The first two contributions, by A. Bonissent and P. Bak, are concerned with equilibrium properties of atomistic systems – liquid or epitaxial layer respectively – in contact with a solid, showing the evolution of spatial order near the solid surface. The subsequent paper by W. Haubenreißer and H. Pfeiffer gives a concise summary of kinetic theories based on lattice models for the solid-fluid interface, todays most powerful approach to describing crystal growth on atomic length scales. The paper by V. V. Voronkov on the phenomenological approach serves as the bridge to the macroscopically observable growth forms, covering length scales from larger than the lattice unit but small compared to surface structures (like spirals) up to the macroscopic shape of the crystal. The article by J. v. d. Eerden finally introduces diffusion as a transport-process, stressing the relative importance of bulk- and surface-diffusion and its influence on surface structures.

In a second volume it is planned to cover a similar series of important developments, starting with equilibrium properties of surfaces of ionic crystals, then reviewing the very recent results on interacting interfaces like surface melting, wetting, layering. The role of dislocations then will be discussed, an extensive review on hydrodynamic flow important for crystal growth will be given and finally the fascinating problems of pattern formation in dendrites and eutectics will be treated.

Our hope is, to draw a sufficiently complete picture of the foreseable development of crystal growth theory. A substantial amount of basic concepts still is waiting for further refinement in order to generally serve as practical guidelines for growing crystals on a production scale in industry. The material presented in this and the following volume, on the other hand, may hopefully act as a stimulus for theories and experiments which will ultimately give us the tools to produce crystals of predefined properties.

A. A. Chernov — Moscow

H. Müller-Krumbhaar — Jülich, 1982

# Table of Contents

# Structure of the Solid-Liquid Interface

**Alain Bonissent**

CRMC2, CNRS, Campus de Luminy, Case 913, F-13288 Marseille-Cedex 9

*Theoretical works on the structure and thermodynamic properties of the solid-liquid interface of a simple substance are reviewed. The methods of investigation follow those which have been applied in the case of the bulk liquids: Bernal random packing of hard sheres, computer simulations and perturbation theory. Application of these techniques allows a description of the interface, in terms of density profile and structure of the interfacial layers. The interfacial specific free energy is estimated in the case of the (111) fcc orientation.*

*Future developments will tend to the calculation of the interfacial free energy in different directions, as well as to a better understanding of the phenomena which occur at the interface during growth of a crystal from the melt.*



Crystals 9

## List of Symbols

| | |
|---|---|
| A | interface area |
| $b_w$ | $\exp[-\beta\phi_w]$ |
| F | Helmholtz free energy |
| $g_0(r)$ | Radial distribution function for hard spheres |
| k | Boltzmann's constant |
| $n_S$, $n_L$ | number of (solid or liquid) molecules |
| $N_m$ | number of molecules in layer m |
| $N^{(i)}_{s,m}$ | number of sites available for molecule i in layer m |
| N | total number of molecules in the system |
| $P(\delta)$ | density profile of the last crystal plane at the interface |
| $s_m$ | configurational entropy per molecule in layer m |
| T | temperature |
| T* | reduced temperature |
| u(r) | intermolecular pair potential |
| $u_1(r)$ | attractive tail of u(r) |
| $U_i$ | potential energy of the system in state i |
| $W_m$ | number of complexions of layer m |
| z | spatial coordinate perpendicular to the solid-liquid interface |
| $z_m$ | value of z for which $\phi_w(z)$ is minimum |
| $\alpha$ | contact angle of a liquid drop on a crystal |
| $\beta$ | 1/kT |
| $\gamma$ | interfacial specific free energy |
| $\varepsilon$ | energy well depth of the intermolecular pair potential |
| $\phi_w$ | potential exerted on the liquid molecules by the crystal face |
| $\phi^r_w$ | repulsive part of $\phi_w$ |
| $\phi^e_w(z)$ | effective wall potential (Boltzmann average of $\phi_w$) |
| $\mu$ | chemical potential |
| $\Xi\{b_w\}$ | grand canonical partition function. The curly brackets denote a functional dependence |
| $\varrho$ | density |
| $\sigma$ | range of intermolecular repulsion |

# I. Introduction

Growth from a melt is extensively used for the preparation of crystals, for mass production (as in metallurgy) as well as for the fabrication of high quality materials necessary for the modern technologies. It is then not surprising that nucleation and growth from the melt have been widely investigated from experimental and theoretical viewpoints. For the growth of crystals, the interface between the two phases plays an important role since it is the region where the molecules are incorporated into the crystal lattice. This is particularly so in the case of growth from a melt where the molecules are always present at the interface. Transformation of liquid molecules into crystalline molecules is in this case essentially a local change in structure, associated with a small density change. The way in which these structural modifications occur is the growth mechanism. which is still not perfectly known for the crystal-melt system.

Before being able to resolve this problem, the structural characteristics of the interface must be determined. In other words, one has to find out how the structure of the liquid is affected by the presence of the crystal face, and also consider the question of crystal symmetry.

One possible assumption is that the liquid behaves like a continuum, and that its density and structure are not affected by the presence of the crystal. In this case, it can be shown that the wetting of the crystal by its own melt would be perfect, or better than perfect. However, a few cases are known of poor wetting, for metals like platinum[1], cadmium[2], bismuth[3] or gallium[4]. These experimental facts show that such a simple assumption is not acceptable, and that the specific structural characteristics of the interface remain to be determined. Several theoretical studies on this problem have been done in the framework of the lattice-like model[5,6,7]: the liquid molecules are supposed to form a lattice, isomorphous to the crystal lattice. The structural differences between the two phases are represented formally by different binding energies between solid and solid, solid and liquid or liquid and liquid molecules respectively. However, if the value of the binding energy between two molecules which belong to the same phase can be obtained from the heat of fusion or vaporization, this is not the case for the binding energy between one liquid and one solid molecule. This can be estimated only if one knows the structure of the interface. Moreover, this model, originally developed for the crystal-vapour interface, does not represent the actual structure of the crystal-liquid interface, and thus does not allow determination of the entropy factors in a realistic way[8].

This chapter is devoted to a presentation of the recent theoretical work dealing with the crystal-melt interface. As we have just seen, the first goal is the determination of the structural characteristics of the interface. These can then be used for the calculation of thermodynamic quantities, which are of experimental interest.

The investigations which will be reported are concerned with the so-called simple substances, which are composed of spherical, chemically inert molecules. Interaction between the molecules is supposed to be pairwise additive (triplets and higher contributions are neglected, as well as the effects of electrons). The pair interaction is assumed to follow some simple law, which gives the strong short-range repulsion, and attraction at intermediate range. The most frequently used intermolecular potential is the well known Lennard-Jones 12–6 potential:

$$u(r) = 4\varepsilon \left[ \left(\frac{\sigma}{r}\right)^{12} - \left(\frac{\sigma}{r}\right)^{6} \right] \quad (1)$$

where r is the distance between the two molecules, $\varepsilon$ is the energy well depth and $\sigma$ is the separation at which $u(r) = 0$.

During the last decade, significant progress has been achieved in this field, expanding upon the previous development of the theories of the liquid state which provided convenient tools for studying high density disordered systems. These techniques can be classified into three groups:

(i) The Bernal model represents the instantaneous structure of the liquid, averaged over the time scale of the local oscillations, so that the distance between two neighbouring molecules is the same all over the model. Towards the time scale of the diffusive movements, it still can be considered as a snapshot. In this approximation, the molecules can be considered as hard spheres. According to Bernal, the liquid is a homogeneous, coherent and essentially irregular assemblage of molecules containing neither a crystalline region nor holes large enough to admit other molecules[9)]. The Bernal model of the simple liquids is a dense random packing of hard spheres with the highest possible density. The statistical characteristics of the geometry of the Bernal model are close to those of the liquid structure, as has been shown by Bernal and Finney[10)]. They also demonstrated the importance of pentagonal symmetry, which is more suitable for high local order, and the non-existence of octahedral clusters, which are typical for the crystalline close-packed structures. Computer-built versions of the Bernal model have been proposed by Bennett[11)] and Matheson[12)]. They consist of a sequential deposition of hard spheres in the tetrahedral sites (pockets) formed by the previously placed spheres.

Application of the ideas of Bernal to the problems of the crystal-liquid interface will be presented in Sect. II.

(ii) The computer simulation techniques have been extensively described elsewhere[13)]. They belong to two classes: the Monte-Carlo[14)] and Molecular Dynamics[15)] techniques. Both techniques apply to systems in which the microscopic parameters are known, namely the intermolecular potential and the molecular mass. A Monte-Carlo simulation produces a Markov chain, i.e. an ensemble of states of the system such that the occurence of a given state, i, is proportional to its thermodynamic probability $\exp(-U_i/KT)$ ($U_i$ is the potential energy of the system in the state i). This is obtained by a random process in which the system is allowed to change from state i to state j with the probability

$$\exp(-\Delta U_{ij}/kT), \text{ where } \Delta U_{ij} = U_j - U_i$$

The molecular dynamics technique consists of the integration of the coupled equations of the movement of the molecules, as defined by Newtonian mechanics. At each time step, the force applied on each molecule by the intermolecular potential is calculated, giving the acceleration of this molecule. Integration over a small time step gives the new position and velocity of each molecule.

In all cases, original conditions must be specified, and an equilibration period must be allowed before the system is in a stable state. Information can be obtained on the structure and thermodynamics of the system being studied. Application to crystal-liquid systems will be described in Sect. III.

(iii) Perturbation theories[16, 17] take advantage of the relative simplicity of the hard-sphere fluid, and of the fact that the realistic intermolecular potentials are sufficiently close to the hard sphere potential. Equations of state for the Lennard-Jones substance have been derived, which compare extremely well with the "experimental" computer-simulated results. Application of these techniques to the crystal-liquid interface is not yet so well developed, although it is a very promising approach. In Sect. IV, we shall describe how the density profile perpendicular to the crystal-liquid interface can be obtained.

## II. Models of the Crystal-Melt Interface

The investigations presented here result from the application of the ideas of Bernal to the crystal-melt interface. The liquid is represented by the hard-sphere Bernal model, in contact with a crystal face. Poor wetting of the crystal by the liquid usually occurs in the case of metals on faces with a hexagonal structure {(111)Fcc or (0001)hcp} and this is why this case has been particularly studied. The two-dimensional hexagonal close-packed lattice has one important peculiarity. It offers to adsorbing atoms of the same size two interpenetrating sublattices of tetrahedral pockets. The occupation of a pocket of one sublattice prohibits the occupation of three neighbouring pockets which belong to the other sublattice. This peculiarity is responsible for the existence of stacking faults in the close-packed crystals. Zell and Mutaftschiev[18] built a model by physically packing some 2000 ping pong balls in a container. The model was then put in contact with a plane with hexagonal structure, composed of spheres of the same size, representing the crystal face. The spheres were then removed one by one, while measuring their coordinates for further analysis. Assuming then a realistic Lennard-Jones potential between the molecules, the potential part of the adhesion energy was calculated. This is equal to the sum of the binding energy of all the "liquid" molecules with all the "solid" molecules (across the interface), reported to an interfacial area unity. The results demonstrate the possibility of a poor wetting of the crystal by the melt. Use of this model was limited, because it did not permit the realization of very large packings.

Another interesting approach was proposed by Spaepen[19]. For Spaepen, the principal characteristic of the liquid structure is the absence of octahedral clusters, which are non-existent in the Bernal model, while abundant in the close-packed crystalline structures. Spaepen assumed that the first liquid layer in contact with the crystal is made of molecules disposed in the tetrahedral sites determined by the crystal molecules, arranged in such a way that the two subsystems of sites are occupied equally (as they correspond to the same potential energy), that the density is maximal (due to the presence of the adjacent liquid phase with a high density), and that no octahedral clusters are formed.

Figure 1 gives an example of the first liquid layer in this model. Geometrical considerations permit an estimation of the density of the first layer, estimated to be about 75% of that of a crystal lattice plane. It is also concluded that there is no density deficit at the interface and thus that the potential part of the interfacial specific free energy is zero. Statistical geometry on the two-dimensional packing at the interface also leads to the entropy of the first liquid layer at the interface, giving access to the interfacial specific

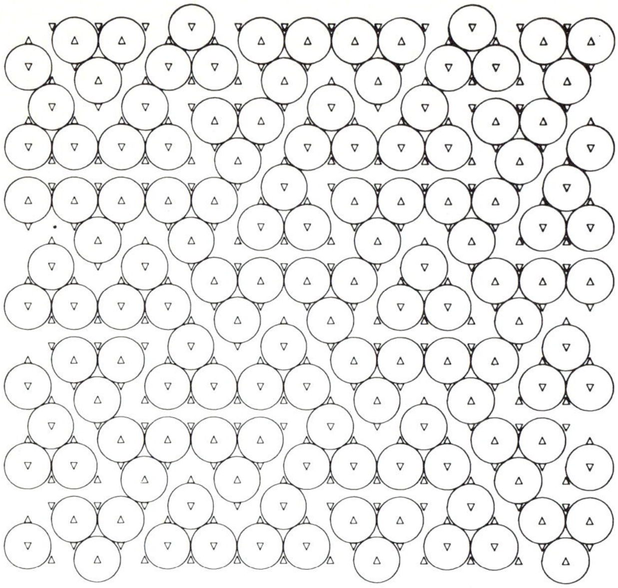

**Fig. 1.** Structure of the first liquid layer in contact with the crystal, as obtained by Spaepen. The two types of triangles △ and ▽ indicate the two interpenetrating sublattices of sites. Reprinted from Ref. 20

free energy. The conclusion was that the crystal is perfectly wetted by the melt. The shortcomings of Spaepen's approach are that the model is limited to one liquid layer and that the basic assumption of non-existence of octahedral clusters at the interface is not justified by thermodynamic considerations.

The last model[20] to be presented here is a computer-built version of the model proposed by Zell and Mutaftschiev[18]. It uses the Matheson[12] algorithm to build a model starting from a crystal face with a hexagonal structure. Like in the Matheson algorithm, the spheres are disposed one by one in the site which is the closest to the original crystal face. If several sites have the same lowest "altitude" z, the site to be occupied is chosen randomly among them. Periodic boundary conditions in the directions parallel to the interface simulate a model with infinite size by elimination of inhomogeneities near the boundaries. Figure 2 shows a cross-sectional view of one of the models obtained. Figure 3 is a representation of the first liquid layer, formed by rafts of spheres in coherent positions (first neighbours), separated by channels in which it is not possible to place a new sphere in contact with the crystal face. This is the beginning of the disorder, which becomes complete in the bulk liquid part of the model, where the structure is similar to that of the Matheson model.

Figure 4 shows the density profile, in the z direction perpendicular to the interface. A slight density minimum can be observed at the interface. This is the origin of the low value of the adhesion energy calculated by Zell and Mutaftschiev[18].

The positions of the molecules, as given by the model, can be used to calculate the interfacial specific free energy if one assumes a realistic Lennard-Jones intermolecular potential[21]. Since we are dealing with two condensed phases with a small density difference, the Gibbs potential or Helmholtz free energy can be used interchangeably. The Helmholtz free energy being directly related to the canonical partition function, we shall

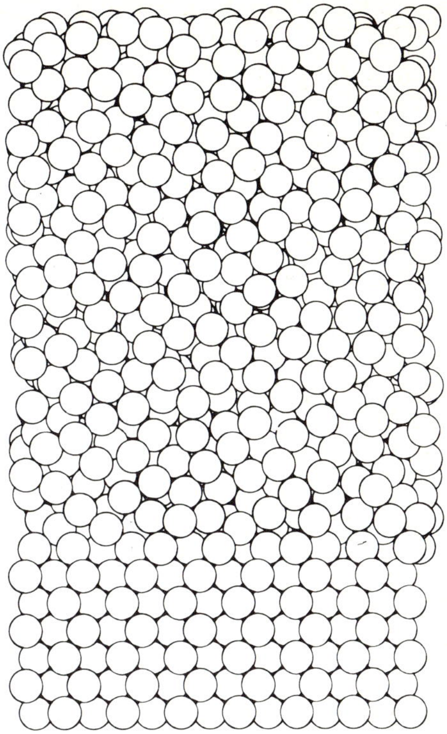

**Fig. 2.** Cross sectional view of a model of the crystal-melt interface. The crystal has an h.c.p. structure and the interface is parallel to the (0001) direction. The plane of the figure is parallel to the $(11\bar{2}0)$ direction. Reprinted from Ref. 21

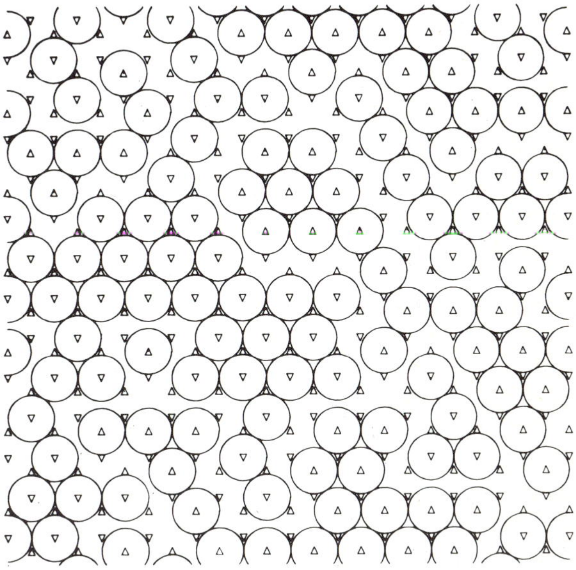

**Fig. 3.** Structure of the first liquid layer in contact with the crystal, in the model of Bonissent and Mutaftschiev. The two types of triangles $\triangle$ and $\triangledown$ indicate the two interpenetrating sublattices of sites. Reprinted from Ref. 20

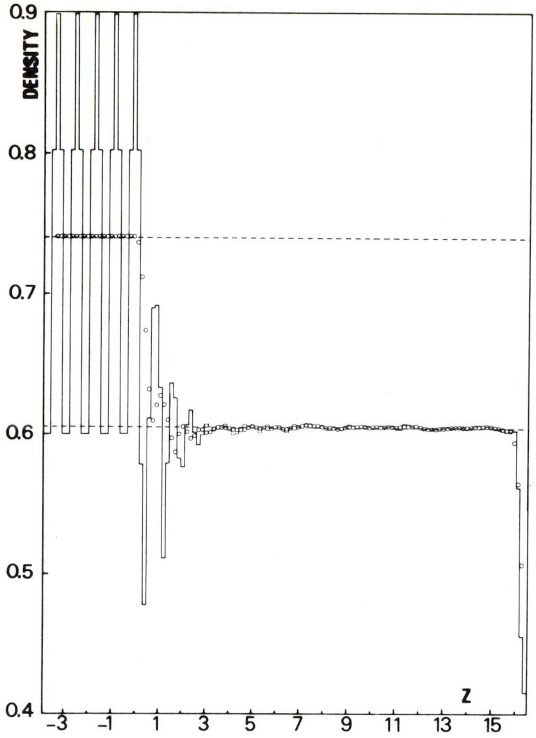


**Fig. 4.** Density profile of a model of the crystal-melt interface, as obtained by Bonissent and Mutaftschiev. The packing fraction is plotted versus the z coordinate perpendicular to the interface. The packing fraction is calculated in slices with thickness 1/5 of the interreticular distance in the crystal. The points (O) represent the local average on five neighboring layers. The interface is oriented in the direction (111) f.c.c. or (0001) h.c.p. From Ref. 21

use it preferentially. The superficial free energy is equal to the difference in energy between a two-phase system and the same number of molecules in the bulk phases:

$$A\gamma_{SL} = F - n_S\mu_S - n_L\mu_L \tag{2}$$

where A is the interface area, $\gamma_{SL}$ is the interfacial free energy, $n_S$ and $n_L$ are the numbers of solid and liquid molecules respectively, and $\mu_S$ and $\mu_L$ are the chemical potentials in the bulk solid and liquid phases.

The problem of the determination of the interfacial free energy from a model of the interface is then, for a given interface area, to determine the free energy, F, of the molecules of the model and the chemical potentials of the bulk phases. This can be done with the model of Bonissent and Mutaftschiev[20]. The free energy F is the sum of a potential and an entropic term. The potential energy is calculated from the positions of the molecules and the intermolecular potential law. The entropy is proportional to the logarithm of the free volume accessible to the molecules. Calculation of the free volume is complex because of the collective character of the movements in the liquid. As a simplification it can be assumed that the molecules undergo oscillatory movements, separated by occasional jumps from one position to another. Therefore, the corresponding configurational integral can be written as the product of local free volumes, and the number of possible space configurations of the liquid molecules, leading to a configurational entropy.

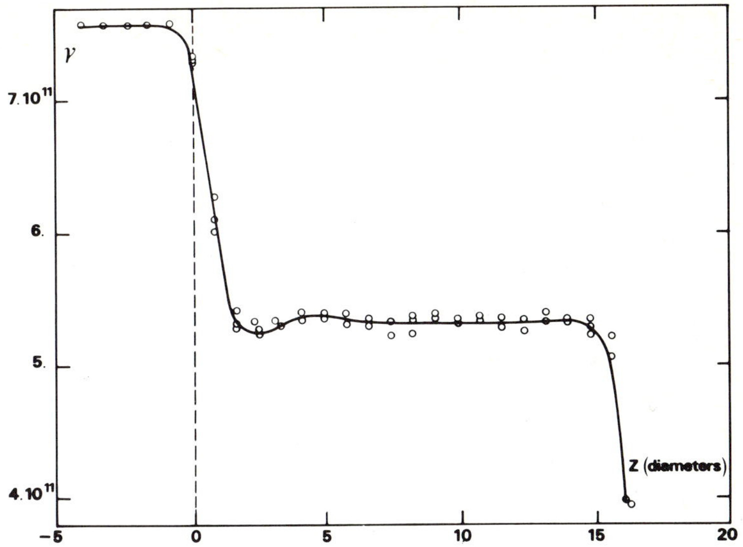


**Fig. 5.** Vibration frequencies, in the Bonissent-Mutaftschiev model of the crystal-melt interface for a molecule, versus its z coordinate, and for argon at the triple point. The *points* (0) correspond to three different models, and the *curve* gives the average values for the three models. From Ref. 21

The vibrational frequencies have been calculated for each molecule in the Einstein approximation, i.e. assuming that the neighboring molecules remain fixed in their equilibrium position. The errors introduced by this approximation should be similar in the bulk and interfacial phases, and are thus minimized. Figure 5 gives the values of the vibrational frequency of a molecule in the model as a function of the z coordinate.

The configurational entropy is estimated during the realization of the models[22]. At any time during the building process, the computer program "knows" all accessible sites. The spheres placed in these sites will form the next ($m^{th}$) "monomolecular" layer. Since the system is disordered, the number of sites is larger than the number of spheres which will form the layer. The number of complexions of this layer is equal to the number of possibilities for occupation of the sites with a maximum number of molecules (this means the highest possible density in the final model). This number is obtained by the random sequential filling method, as proposed originally by Baker[23] for the two-dimensional honeycomb lattice. At a given moment of the building process, a random sequential filling is performed. All present sites are accounted for by allowing them to be occupied randomly and removing, after deposition of a new sphere, all sites whose occupation becomes impossible. The sites created thereby are not taken into consideration. After deposition of some $N_m$ spheres, a "jamming" occurs: it becomes impossible to place a new sphere, since there is no site available. The number of possibilities for placing each sphere (labelled i) is identical to the number $N_{s,m}^{(i)}$ of disposable sites. Since the spheres (molecules) are undistinguishable, the number of complexions of the layer is:

$$W_m = \left(\prod_{i=1}^{N_m} N_{s,m}^{(i)}\right)/N_m! \,, \tag{3}$$

and the entropy per molecule is:

$$S_m = \frac{1}{N_m} k \ln W_m . \tag{4}$$

Figure 6 presents the configurational entropy per molecule, s, at various distances from the interface.

The estimate of the specific crystal-melt interfacial free energy for Argon at the triple point, by the methods outlined above, is:

$$\gamma_{SL} = 11.2 \text{ mJ/m}^2 , \tag{5}$$

Applying the same methods, the crystal-vapor and liquid-vapor interfacial energies are estimated, for the same substance, respectively as

$$\gamma_S = 32.6 \text{ mJ/m}^2 , \tag{6}$$

$$\gamma_L = 22.1 \text{ mJ/m}^2 , \tag{7}$$

These are calculated assuming that the liquid and crystal structure and density are not affected up to the liquid-vapour or crystal-vapour interfaces.

The contact angle $\alpha$ for a liquid droplet on a (111) face of solid argon is determined by:

$$\cos \alpha = \frac{\gamma_S - \gamma_{SL}}{\gamma_L} = 0.97 , \tag{8}$$

$$\alpha = 14° . \tag{9}$$

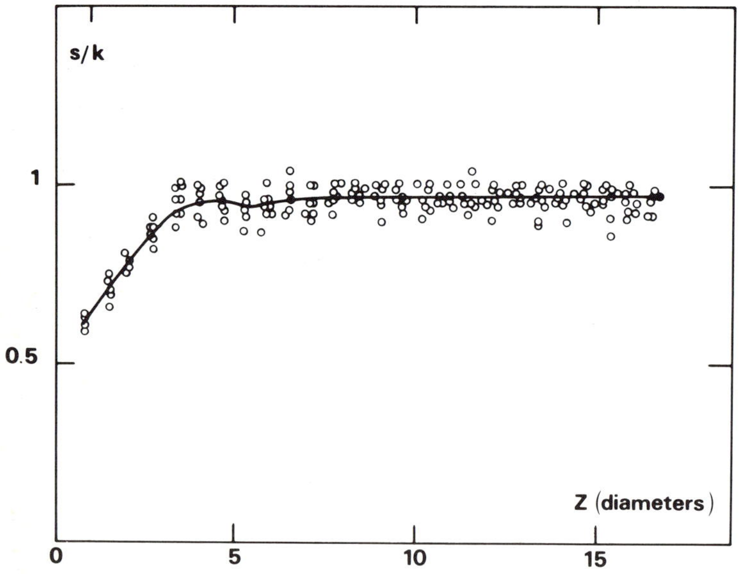


**Fig. 6.** Configurational entropy per molecule as a function of the z coordinate, as calculated during the building process. The *points* (O) correspond to 12 different models. The *solid line* represents the average value at each point. From Ref. 22

This is in qualitative agreement with the experimental finding of poor wetting of some dense faces of crystals by their own melt[1–4]. The limitations of this approach are those of the Bernal model, namely that the effective intermolecular potential is not taken into account during the building process, and that the thermal motion does not appear explicitly. In particular, the thermal motion of the molecules in the crystal can cause a deformation of the interface, or some kind of surface roughness, the character of which remains to be determined. It is then of great interest to check (by another method) the structure of the liquid near the crystal face as determined by the models. This is the object of the following section.

## III. Computer Simulations

Improvement in the capabilities of computers made it possible to apply simulation techniques to study the crystal-melt interface of a simple substance.

The first simulation of a system containing a crystal-melt interface was that of a three-phase system, as reported by Ladd and Woodcock[24]. This was a molecular dynamics simulation of a Lennard-Jones substance. The system, consisting of 1500 molecules, was limited by periodic boundaries in two (x-y) directions. In the z direction, it was composed consecutively of a static crystal, with thickness larger than the effective range of the interatomic potential, of a mobile crystal slab, of a liquid slab and finally of a vapor up to a wall. The crystal has an f.c.c. structure, and the interface is oriented in the (100) direction. The initial conditions and the equilibration period are manipulated in order to obtain this configuration at the desired triple point reduced temperature. Because of the presence of the static crystal, the liquid and solid phases separate during equilibration, and the system exhibits crystal-liquid and liquid-vapor interfaces.

The main conclusion of this work is that the interface is rather diffuse, since the density profile exhibits a layered structure in the liquid phase, over some 5 atomic

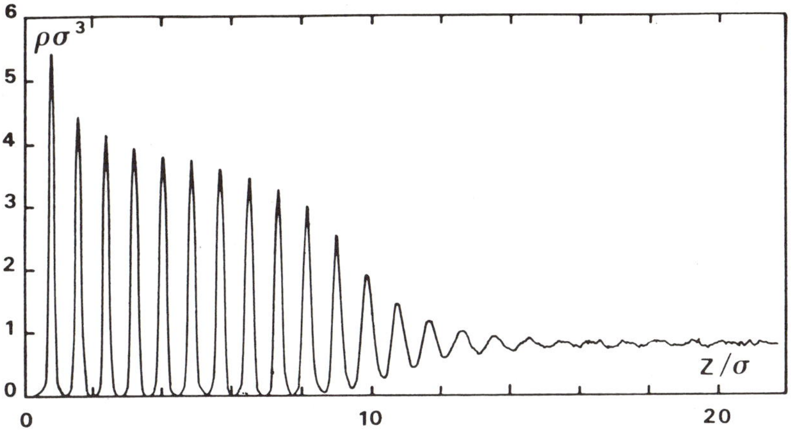


**Fig. 7.** One particle density profile obtained by Ladd and Woodcock[26] for the (100) crystal-melt interface of a Lennard-Jones substance

diameters, as shown in Fig. 7. The exact character of that ordering was not determined. The density profile of the liquid-vapor interface is also obtained, and appears to be consistent with other simulations.

Another simulation of a system involving a crystal-melt interface was reported in the same period by Toxvaerd and Praestgaard[25]. It consists of a molecular dynamics simulation on a set of some 1700 Lennard-Jones molecules. The system consists of a "sandwich", composed of a liquid slab enclosed between two solid slabs, thus defining two solid-liquid interfaces. The system is surrounded by periodic boundaries in all three directions. During the equilibration period, the z extension of the system (perpendicular to the interfaces) is adjusted, until the pressure is that of equilibrium at the chosen reduced temperature $T^* = 1.15$. The temperature is adjusted by the usual rescaling of the velocities. The interface is oriented in the (100) direction of the (f.c.c.) crystalline part of the system. The conclusions are similar to those of Ladd and Woodcock, in particular that the density profile perpendicular to the interface exhibits oscillations over some 5 atomic diameters.

The main interest of these two investigations was to show that it is possible to keep a two-phase system in equilibrium long enough to observe the properties of the crystal-melt interface. This was used in subsequent work on the subject.

In a second paper, Ladd and Woodcock[26] performed a more precise analysis on their system, during a long run after equilibration. Special attention was given to the density profile, potential energy profile, diffusion coefficients and trajectories of the molecules. It was concluded that the interface extends over several atomic diameters, and that the physical properties change gradually from those of the crystal to those of the liquid, across the interface. This differs from the conclusions drawn from the hard sphere models described in the previous section. However, two different faces were concerned ((111) versus (100) orientations), which may cause some of the differences.

The purpose of the simulation by Bonissent, Gauthier and Finney[27] was to check the validity of the assumptions on which the hard sphere model presented in the previous section is based. Comparison with a computer simulated soft sphere system should be illuminating. The computer experiment[27] consists of a Monte-Carlo simulation on a system consisting of some 860 Lennard-Jones molecules. Periodic boundary conditions are again imposed in the x-y directions. In the z direction, there are successively a static crystal, a mobile crystal slab of two reticular planes, a liquid slab and finally the vapor phase. The interface has the (111) orientation and the starting configuration is realized by the hard sphere sequential deposition technique described in Sect. II. The hard sphere potential is then replaced by a Lennard-Jones potential. The transitional period of thermal equilibration, over which the potential energy changes significantly is short, confirming that the structure of the soft sphere system is close to that of the hard sphere model. The potential energy difference is evaluated to 5% of the total potential energy.

The density deficit observed in the hard sphere model remains, although of smaller amplitude, and the extension of the interfacial zone is larger (about 5 atomic diameters in the soft sphere system).

Observation of the trajectories of the molecules shows that the first liquid layer, on which the island-like structure remains, has essentially the behavior of a crystal plane with a big concentration of defects. The next (liquid) layer behaves like a liquid, as can be deduced from the observation of the mean square displacement versus computer time, for molecules situated at different positions in the system. It is not clear, however, to what extent these properties are affected by the small extension of the mobile crystal in the z direction.

The last contribution to be presented here is a comparative molecular dynamics simulation of two Lennard-Jones systems containing solid-liquid interfaces, with the (100) and (111) orientations respectively. In both cases, the system consists of a liquid slab between two crystal slabs, and limited by periodic boundaries in the three directions. The interfaces are perpendicular to the z direction. In the starting configuration, the (future) liquid part is formed by a z expanded crystal, whose density is that of the triple point liquid, as determined by Ladd and Woodcock[24]. The temperature is that of the triple point, $T^* = 0.67$.

Figure 8 shows the trajectories of the molecules over 5000 time steps performed at equilibrium, for both systems.

Figure 9 and 10 present the same movements in cross-sectional views perpendicular to the plane of Fig. 9. They show that the two systems appear to behave very similarly. Layer 6 of the (111) system and layer 7 of the (100) system can be considered as equivalent. They present a quasi-crystalline layer, in which the amount of diffusion is not negligible, although the molecules spend most of their time in the lattice sites. The next layer up, in each case, represents a two-dimensional liquid, but careful examination of the trajectories indicates a higher residence time in the positions of the ideal lattice sites. This is due to the potential minima created by the field of the layer underneath, which is practically crystalline. The two-dimensional radial distribution functions and the diffusion coefficients calculated for the molecules of the same slices confirm these observations. We can then conclude that as far as the structural characteristics are concerned, the interface between the crystal and the melt extends essentially over two atomic diameters. The subsequent oscillations in density or related quantities, observed in this simulation as well as in others, are minor compared with the abrupt change of symmetry wich occurs at the interface itself. This is in good agreement with the conclusions obtained from the hard sphere models of the (111) interface.

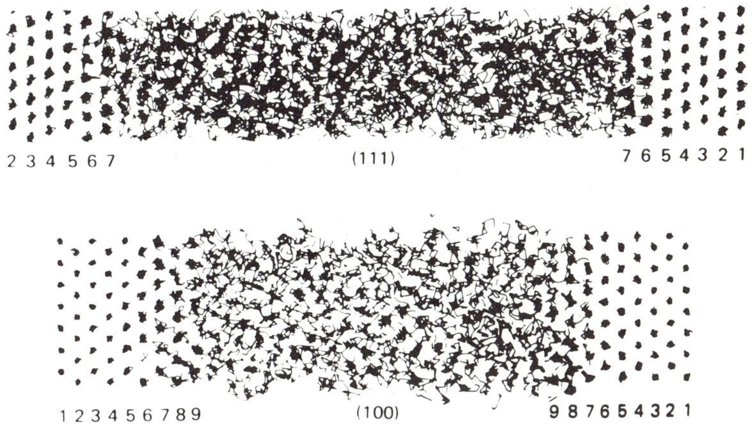


**Fig. 8a, b. Trajectories of the molecules during the simulation in a slice perpendicular to the interface (x-z plane) for the (a)** (111) and **(b)** (100) systems. Any atom entering the slice, at any time during the simulation, is represented so long as it remains in the slice. From Ref. 28

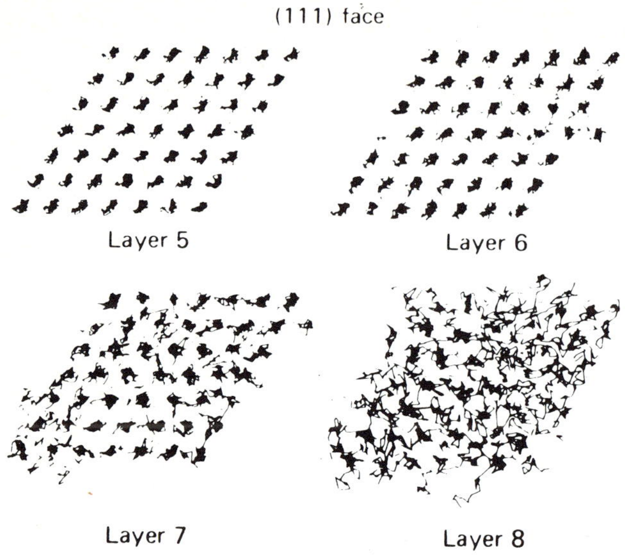


**Fig. 9.** Trajectories of the molecules in layers parallel to the interface (x-y plane) using the same procedure as for Fig. 10, (111) interface. From Ref. 28

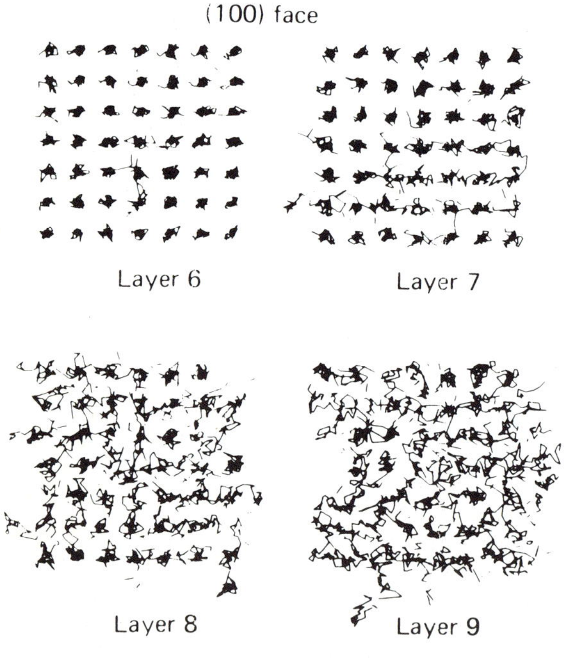


**Fig. 10.** Trajectories of the molecules in layers parallel to the interface (x-y plane), (100) interface. From Ref.28

Since the potential energy of the system is known at each moment during the simulation, the potential part of the interfacial free energy can be obtained. The local free volume, or vibrational entropy, is determined from the mean free path of the molecules during the simulation. However, the configurational entropy cannot be obtained by this method. Therefore, the interfacial free energy cannot be calculated.

Computer simulation techniques are a powerful tool for studying the crystal-melt system, as they allow simulation of interfaces with any orientation, for realistic Lennard-Jones systems. It is infortunate that these methods do not allow calculation of the free energy of the systems studied. Another drawback is the need for long computer runs, involving non-negligible costs.

## IV. Perturbation Theory

Perturbation theories are presently used to predict the properties of bulk liquids. The reference system is the hard sphere fluid, for which a satisfying representation is obtained in the Percus-Yevick[41)] approximation, or from computer experiments. Even if computer simulated results are used, the simplification of the algorithms for the hard sphere potential and the universal character of the reference system justify the use of the perturbation theory. Perturbation techniques have been applied to the liquid-vapor interface[29)]. The density profile and surface tension are obtained with very good accuracy for the Lennard-Jones liquid. The case of the solid-liquid interfaces is more complicated since, as we have shown in the previous sections, the interface is characterised by a special structure, which is not entirely represented by the density profile perpendicular to the interface. This is why it has not yet been possible to calculate the interfacial specific free energy. A treatment for the density profile is known. This is the Abraham-Singh perturbation theory for non-uniform fluids[30)] applied to the case of the crystal-melt interface[31)].

The Abraham-Singh theory is inspired by the Weeks-Chandler-Andersen[17)] perturbation theory. In the latter, it has been shown that a fluid with a purely repulsive soft potential has the same properties as a hard sphere fluid, if the hard sphere diameter is chosen adequately. The effect of the attractive tail of the intermolecular potential is introduced by the perturbation method. It is to first order:

$$\Delta F = \frac{1}{2} N\varrho \int u_1(r)\, g_0(r)\, d\vec{r} \ ,$$

where N is the number of molecules, $\varrho$ is the density, $u_1(r)$ the attractive part of the potential, and $g_0(r)$ the radial distribution function of the hard sphere fluid. The higher order terms are small and can be neglected.

In the Abraham-Singh theory, it is assumed that the interaction potential $\phi_w$ between the solid and any atom in the fluid depends only on the normal distance z between the solid face and the fluid atom. It is also assumed that the structure of the fluid at the interface is determined by the repulsive nature of the potential $\phi_w^r(z)$. Accordingly, the potential exerted on the liquid, due to the presence of the solid face, is considered to be that of a soft repulsive wall:

$$\begin{aligned} \phi_w^r(z) &= \phi_w(z) - \phi_w(z_m) \ , \qquad && z < z_m \\ \phi_w^r(z) &= 0 \ , && z > z_m \end{aligned} \qquad (7)$$

$Z_m$ is the position at which $\phi_w(z)$ is minimum.

The grand canonical partition function, $\Xi$, of the fluid-wall system is treated as a functional of the Boltzmann factor $b_w^r(z) = \exp[-\beta\phi_w^r(z)](\beta = 1/kT)$, and a Taylor expansion of $\ln \Xi\{b_w^r\}$ is performed in powers of the Andersen-Weeks-Chandler "blip" function, $\Delta b_w^r(z) = b_w^r(z) - b_w^h(z)$, where $b_w^h(z) = \exp[-\beta\phi_w^h(z)]$ is the Heaviside step function for a hard wall.

The expansion is

$$\ln \Xi\{b_w^r\} = \ln \Xi\{b_w^r\} + \int y_w^h(z)\Delta b_w(z)dz + \dots \tag{10}$$

where

$$y_w^h(z) = \delta \ln \Xi\{b_w^h\}/\delta b_w^h(z) = \varrho^h(z)\exp[\beta\phi_w^h(z)] \tag{11}$$

$\varrho^h(z)$ is the single particle density distribution of the fluid in contact with a hard wall. The position of the hard wall is chosen so that the second term in Eq. (10) vanishes:

$$\int y_w^h(z)\Delta b_w(z)dz = 0\ . \tag{12}$$

Therefore,

$$\ln \Xi\{b_w^r\} \simeq \ln \Xi\{b_w^h\} \tag{13}$$

Equation (13) can be used to determine the density distribution $\varrho^r(z)$ of the fluid in contact with the soft repulsive wall. Abraham and Singh find that:

$$\varrho^r(z) \simeq \exp[-\beta\phi_w^r(z)]y_w^h(z) \tag{14}$$

Equations 12 and 14 are the basic results of the Abraham-Singh theory. The values of $y_w^h(z)$ for the hard sphere fluid in contact with a hard wall have been obtained by Henderson, Abraham and Barker[32)], by applying the Percus-Yevick approximation to a mixture of spheres with two diameters, in which one of the components becomes infinitely dilute and infinitely large in size. The validity of the Abraham-Singh theory has been established by comparing it with Monte-Carlo simulations of a hard sphere fluid in contact with a repulsive Lennard-Jones 9-3wall[30)].

When applying the Abraham-Singh theory to the crystal-melt interface, two questions arise which are not relevant when the idealized soft wall system is considered. These are the effect of the two-dimensional structure of the potential exerted by the solid on the liquid molecules and the thermal motion of the crystal molecules. Due to the periodicity of the crystal lattice, it is useful to expand the potential $\phi_w(x, y, z)$ in a two-dimensional Fourier series in the x-y plane. The leading term is a function of the z coordinate only. In the present state of the theory, this is the only term which can be taken into account. The higher order terms have to be neglected. Abraham[33)] has shown that if the expansion is performed not on the potential, but on $\exp[-\phi_w(x, y, z)/kT]$, the leading term is a more accurate approximation of the exact potential. This is due to the fact that the Boltzman factor of the wall potential plays a dominant role in the statistical mechanics of the system, through the expression of the partition function. Another reason is that the exp-function is bounded by 0 and 1, while the $\phi$-function has very sharp peaks for a molecule situated in the neighborhood of the crystal face. Thus, the exp-function is more regular and will be better approximated by its average value, which is the leading term in the Fourier series expansion. Following Abraham, the effective potential $\phi_w^e(z)$ is defined by:

$$\exp[-\phi_w^e(z)/kT] = \frac{1}{A}\int_A \exp[-\phi_w(x, y, z)/kT]dxdy \tag{15}$$

The second point concerns the thermal motion of the surface atoms. In a crystal face, it can be assumed that neighboring atoms perform "in phase" oscillations, i.e. it is unlikely that the positions of any two, three or four neighboring surface atoms would be highly uncorrelated. Furthermore, the liquid structure near the crystal face is determined primarily by the configurational distribution of the molecules in the surface sites of the crystal face. We can also consider that, to a good approximation, the density profile in the direction normal to a crystal plane is dominated by the vibrational modes with a wavelength significantly larger than the crystal lattice parameter (i.e., those for which an adsorption site at the interface is distorted very little). From this viewpoint, the interfacial density profile $\varrho^T(z)$ at temperature T is altered from the zero-temperature crystal-fluid profile using the following crude approximation:

$$\varrho^T(z) = \int \varrho^O(z + \delta)P(\delta)\, d\delta / \int P(\delta)\, d\delta \tag{16}$$

where $P(\delta)$ is the density profile in the normal direction of the crystal face. $P(\delta)$ can be obtained from the computer simulation results. Finally, the hard sphere diameter for the liquid must be adjusted according to the interatomic potential, temperature and density, for the system under consideration. This is done using the Weeks-Chandler-Andersen theory[17)].

The results are given in Fig. 11 and 12 for the (111) and (100) faces respectively, and compared with the molecular dynamics results. We see that the agreement is rather good, especially if one recalls the crudeness of the approximations used. This may be due to the fact that the function considered $\varrho(z)$ is a two-dimensional average, thus having a weak dependence on the detailed two-dimensional structure of the crystal face. The agreement is, however, much better on the (111) face than on the (100) face. The reason is that the reference system is the same (hard spheres against a hard wall) in the two cases.

It has been shown by Bernal[9)] that the structure of a hard sphere packing near a flat wall exhibits hexagonal symmetry. This is obviously closer to the (111) interface than to the (100). Since application of the perturbation theory modifies only the first peak of the density profile (the one which is in the range of the wall potential), it is not surprising to observe a good fit only for this peak in the (100) interface. The subsequent peaks remain those of a hard sphere fluid against a hard wall, and are more representative of the structure of the (111) interface than of the (100) interface.

Another perturbation theory has been proposed by Haymet and Oxtoby[34)]. In their treatment, the bulk liquid is the reference system and the effect of the nonuniformity is treated by using an effective one-body potential which depends on the environment of the molecule considered. This should be an adequate formalism, since one is not concerned in this case with the properties of the interface itself, but rather by the differences between the interface and the bulk material which determine the interfacial free energy.

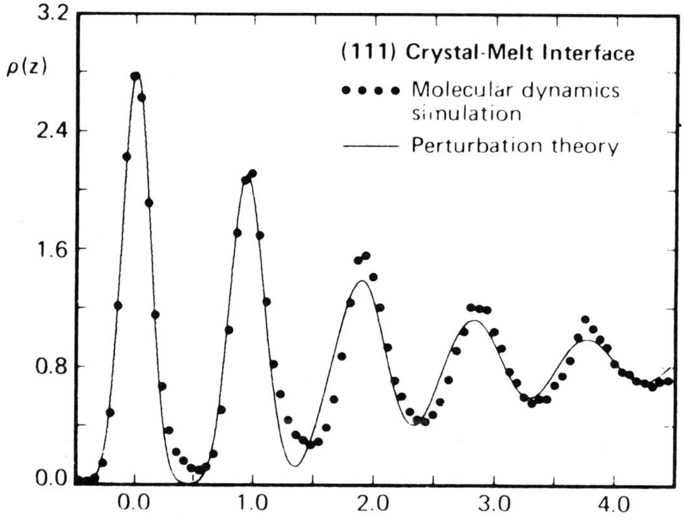


**Fig. 11.** Density profile of the melt neighboring the (111) face predicted by the perturbation theory and compared with the "experimental" profile obtained by molecular dynamics simulation. From Ref.28

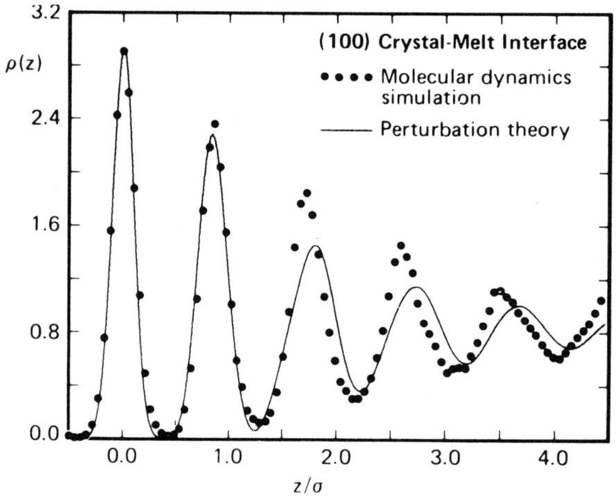


**Fig. 12.** Same as Fig. 11, for the (100) face

This treatment expands upon a previous work by Ramakrishnan and Youssouff[35], which represents the uniform crystal as a perturbed uniform liquid. The density change and the magnitude of the Fourier components of the expansion of the local density under freezing are obtained. The case of the crystal-melt interface is treated by Haymet and Oxtoby in a similar manner, leading to expressions for the interfacial free energy and interfacial density profile. It remains to check the accuracy of this theory by comparison with the results obtained by other methods.

## V. Conclusion

Each one of the methods presented gives a limited amount of information on the structure and related thermodynamic properties of the interface. In this respect, they complement each other for a study of the crystal-melt system. In conclusion, it is interesting to summarise the main results which have been obtained by each one of the techniques used.

Up to now, the model approach is the only method which permits an estimation of the interfacial specific free energy, as the algorithm used gives access to the configurational entropy. The results obtained predict a poor wetting of the (111) crystal face by the liquid. Since the model approach involves approximations whose validity was demonstrated by Bernal and co-workers only in the case of the bulk liquids, it is useful to compare the structure of the models with that of the "experimental" computer-simulated systems.

(i) The computer simulations show that, for the (111) as well as for the (100) faces, the interface is composed essentially of two layers, one of them being quasi-crystalline and the other one being quasi-liquid. The effective range on which the structure is perturbed by the presence of the crystal face extends over 5–6 atom diameters, but the important structural changes occur abruptly in the two layers mentioned. This is in good agreement with the structure of the hard sphere models of the (111) interface. The characteristic island-like structure of the last quasi-crystalline layer, observed in hard sphere models, is not visible in the computer simulations. This can be due to the thermal motion of the molecules in the soft spheres system. A consequence of these movements is that, although this is its average position (this can be seen on the pictures of the trajectories), a molecule is never exactly in the corresponding pocket, which makes the structural characterization more difficult. The same reason might be the origin of the non-existence of the density deficit, which is observed in the hard sphere models. The oscillation of the surface crystalline atoms as introduced in Sect. IV, has the effect of spreading the density peaks, which become wider and of smaller amplitude. The small density minimum might thus disappear. It is indicative that in the Monte-Carlo simulation of Bonissent et al.[27], the minimum is present, with a smaller amplitude than in the hard sphere model. Remember that in this simulation, only two crystalline planes at the interface are free to move. The surface molecules then perform movements of smaller amplitude than in the more realistic simulations.

(ii) The direct approach with perturbation theory does not bring any new results, since density profiles are already known from computer simulations. It must be considered instead as a self-teaching process, in the sense that it provides information on the validity of the approximations and thus on the physics of the system.

From this review, we see that the knowledge of the crystal-melt interface is far less advanced than that of the crystal-vapor or liquid-vapor interfaces. This is related to the presence of the two adjacent condensed phases, from which structural peculiarities occur. As far as crystal growth is concerned, the basic questions like: what is the growth mechanism or how is the crystal nucleus formed in the case of nucleation from the melt[36, 37] are still unanswered. A better knowledge of the structural and thermodynamic properties of the crystal-melt interfaces will shed light on these problems.

For further developments, all the three approaches presented are equally promising but each suffers from deficiencies. The model approach is unfortunately limited to the

(111) face, and the computer simulations have the inconvenience that the computer time necessary to perform the simulations is costly. These limitations will probably be eliminated, either by more skill on the part of the researchers, or by computers with higher capabilities and lower costs. The perturbation techniques can help by allowing efforts to be concentrated on more simplified systems. The cell theories[38], which had been extensively used for the bulk liquids, have not been applied to the crystal-melt interface. A simple reason is that none of these theories was found to describe the liquid correctly. This might change in the future since a new, self-consistent, cell theory has been proposed by Barker[39, 40], which provides equations for calculating the properties of either the solid or the liquid phases. This formalism is thus particularly adequate for studying a two phase system.

## VI. References


1. Kaischew, R., Mutaftschiev, B.: Z. Phys. Chemie *204,* 334 (1955)
2. Stranski, I. N., Paped, E. K.: Z. Phys. Chemie *B38,* 451 (1937)
3. Glicksman, M. E., Vold, C. L.: Acta Met. *17,* 1 (1969)
4. Vollmer, M., Schmidt, O.: Z. Phys. Chem. Leipzig, *467* (1937)
5. Jackson, K. A., Uhlmann, D. R., Hunt, J.: J. Cryst. Growth *1,* 1 (1967)
6. Mutaftschiev, B.: In: Adsorption et Croissance Cristalline, p. 231 CNRS, Paris 1965
7. Temkin, D. E.: In: Crystallization Processes, p. 15. Consultant Bureau, New York 1966
8. Mutaftschiev, B.: Materials Chemistry *4,* 263 (1979)
9. Bernal, J. D.: Proc. Roy. Soc. *A 280,* 299 (1964)
10. Finney, J. L.: Proc. Roy. Soc. *A 319,* 495 (1970)
11. Bennett, C. H.: J. Appl. Phys. *43,* 2727 (1972)
12. Matheson, A. J.: J. Phys. C.: Solid State Physics *7,* 2569 (1974)
13. Hansen, J. P., McDonald, I. R.: Theory of Simple Liquids, Academic Press, London, New York, San Francisco 1976
14. Metropolis, N. et al.: J. Chem. Phys. *21,* 1087 (1953)
15. Alder, B. J., Wainwright, T. E.: J. Chem. Phys. *27,* 1208 1957)
16. Barker, J. A., Henderson, D.: J. Chem. Phys. *47,* 4714 (1967)
17. Andersen, J. C., Weeks, J. D., Chandler, D.: Phys. Rev. *A 4,* 1597 (1971)
18. Zell, J., Mutaftschiev, B.: J. Cryst. Growth *13/14,* 231 (1972)
19. Spaepen, F.: Acta Met. *23,* 731 (1975)
20. Bonissent, A., Mutaftschiev, B.: Phil. Mag. *35,* 65 (1977)
21. Bonissent, A.: These d'Etat, Marseille (1978)
22. Bonissent, A., Finney, J. L., Mutaftschiev, B.: Phil. Mag. *B 42,* 233 (1980)
23. Baker, B. G.: J. Chem. Phys. *45,* 2694 (1966)
24. Ladd, A. J. C., Woodcock, L. V.: Chem. Phys. Letters *51,* 155 (1977)
25. Toxvaerd, S., Praestgaard, E.: J. Chem. Phys. *11,* 5291 (1977)
26. Ladd, A. J. C., Woodcock, L. V.: J. Phys. C.: Solid State Physics *11,* 3565 (1978)
27. Bonissent, A., Gauthier, E., Finney, J. L.: Phil. Mag. *B 39,* 49 (1979)
28. Broughton, J. Q., Bonissent, A., Abraham, F. F.: J. Chem. Phys. *74,* 4029 (1981)
29. Singh, Y., Abraham, F. F.: J. Chem. Phys. *67,* 537 (1977)
30. Abraham, F. F., Singh, Y.: J. Chem. Phys. *67,* 2384 (1977)
31. Bonissent, A., Abraham, F. F.: J. Chem. Phys. *74,* 1306 (1981)
32. Henderson, D., Abraham, F. F., Barker, J. A.: Mol. Phys. *31,* 1291 (1976)
33. Abraham, F. F.: J. Chem. Phys. *68,* 3713 (1978)
34. Haymet, A. D. J., Oxtoby, D. W.: J. Chem. Phys. to be published
35. Ramakrishnan, T. V., Youssouff, M.: Phys. Rev. *B 19,* 2775 (1979)
36. Hsu, C. S., Rahman, A.: J. Chem. Phys. *71,* 4974 (1979)

37. Wendt, H. R., Abraham, F. F.: Phys. Rev. Letters *41,* 1244 (1978)
38. Barker, J. A.: in: Lattice Theories of the Liquid State, Pergamon Press, Oxford, London, New York, Paris 1963
39. Barker, J. A., Gladney, H. M.: J. Chem. Phys. *63,* 3870 (1975)
40. Cowley, E. R., Barker, J. A.: J. Chem. Phys. *73,* 3452 (1980)
41. Percus, J. L., Yevick, G. J.: Phys. Rev. *110,* 1 (1958)

# Melting and Solidification of Epitaxial Structures and Intergrowth Compounds

**Per Bak**

H. C. Ørsted Institute, Universitetsparken 5, DK-2100 Copenhagen Ø, Denmark

*In this chapter theories for melting and solidification of epitaxial structures and intergrowth compounds are reviewed. It is a common feature of these systems that the melting of one lattice takes place in the presence of another "host" lattice or substrate, so the solidification does not start from a fully rotationally and translationally symmetric liquid phase. A number of unusual and interesting situations may occur. For instance, continuous melting may take place in both two and three dimensions. In this case, solidification is not expected to start from nucleation and proceed through crystal growth, but will rather occur throughout the system in a more uniform way, with precursor effects characteristic for second-order transitions.*

*A variety of physical systems will be reviewed. These include rare-gas monolayers adsorbed on graphite, metal surfaces undergoing reconstruction transitions, graphite intercalation compounds, and mercury chain compounds. Several types of transitions are encountered: Potts transitions in two and three dimensions, XY-transitions, Nelson-Halperin transitions, Ising transitions, and transitions described by more esoteric n-vector models.*



Crystals 9
Springer-Verlag Berlin Heidelberg 1983

# 1 Introduction

The transition between a high-temperature liquid phase and a low temperature solid phase is almost always first-order (discontinuous). There is a latent heat involved in the transition, and there is an assymmetry between the transition from liquid to solid ("crystal-growth") and the transition from solid to liquid. Such hysteresis is characteristic for first order transitions. For instance, it is possible to supercool a liquid, but not to superheat the solid. The transition from liquid to solid takes place at a temperature which is lower than the equilibrium transition temperature where the free energies of the two phases are the same, without any precursor effects. The solidification transition can thus not be described by equilibrium thermodynamics, but it is necessary to study the non-equilibrium dynamics of the system. Nucleation and crystal-growth are the dynamical mechanisms by which the solidification takes place in most systems.

This chapter deals with the situation where the solidification takes place in the presence of another lattice, the "host" lattice, or a substrate, for instance a graphite crystal. In such cases the process does not start from a fully rotationally and translationally symmetric liquid phase. The host lattice, which does not itself undergo any transition, breaks the symmetry of the liquid phase. We shall see that this may alter the melting and solidification processes dramatically. For instance, in some cases the transition may become *continuous* and reversible. The transition may then be described by means of traditional equilibrium thermodynamics, and there is a less urgent need for studying the dynamics. For continuous transitions one expects strong precursor effects (fluctuations) near the transition. The specific heat, an certain mass-density correlation lengths diverge as the transition is approached.

The physical systems that will be reviewed include rare-gas monolayers physisorbed on graphite, metal surfaces undergoing reconstruction transitions, graphite intercalation compounds, and the mercury chain compound $Hg_{3-\delta}AsF_6$. A number of different types of transitions appear in these materials: Potts transitions in two and three dimensions, XY-transitions, Nelson-Halperin transitions, melting of discommensuration networks, and transitions described by more complicated n-vector models. But before going to the specific cases, let us briefly consider the question:

## 1.1 Why is Melting Usually a First-Order Transition?

It is quite easy to see that the discontinuous nature of the melting transition can be understood as a consequence of the full rotational symmetry of the liquid phase. Let $\varrho(\vec{r})$ denote the thermodynamic average mass density of the system. In the liquid phase $\varrho(\vec{r})$ is a uniform function

$$\varrho(\vec{r}) = \varrho_0 ,$$

but in the solid phase the formation of a lattice obviously involves an excess mass density with the symmetry of the lattice

$$\varrho(\vec{r}) = \varrho_0 + \delta\varrho(\vec{r}) .$$

The density $\delta\varrho$ can be Fourier transformed,

$$\delta\varrho(\vec{r}) = \sum_{\vec{k}} \varrho_{\vec{k}} \exp(i\vec{k}\cdot\vec{r}) , \tag{1.1}$$

i.e. the solidification can be described as the formation of mass-density waves. Now, let us consider the Landau expanion[1] of the free energy

$$F = \sum_{\vec{k}} A_{\vec{k}} \varrho_{\vec{k}} \varrho_{-\vec{k}} + F_{3+} \dots . \tag{1.2}$$

Because of rotational symmetry $A_k$ can depend only on the magnitude of $\vec{k}$. As the transition is approached, $A_k$ becomes smaller and approaches zero for the most favorable wave-vector $|\vec{k}_1|$ which minimizes $A_k$, and eventually it becomes favorable to form a lattice with this vector as a reciprocal lattice vector.

The important point is that it is possible to form third order terms in the expansion using only wave-vectors of the same length $|k_1|$:

$$F_3 = B_{\vec{k}_1} \varrho_{\vec{k}_1} \varrho_{\vec{k}_2} \varrho_{\vec{k}_3} \delta(\vec{k}_1 + \vec{k}_2 + \vec{k}_3) . \tag{1.3}$$

The third order term is translationally invariant if $\vec{k}_1$, $\vec{k}_2$ and $\vec{k}_3$ form an equilateral triangle (Fig. 1). In two dimensions the "triple $\vec{k}$" structure corresponds to a triangular lattice, in three dimensions to a rod-like structure. Alexander and McTague[2] pointed out that the third order term would be even smaller if *six* pairs of wave-vectors $\pm\vec{k}_i$ are combined to form an octahedron (Fig. 1). The superposition of the six mass-density waves gives a structure of BCC symmetry. The free energy may be expanded in terms of the order parameter $\varrho$, where

$$\varrho^2 = \sum_{n=1}^{6} \varrho_{\vec{k}_n}^2$$

and takes the form

$$F = A\varrho^2 - \frac{2B}{3\sqrt{3}} \varrho^3 + \dots . \tag{1.4}$$

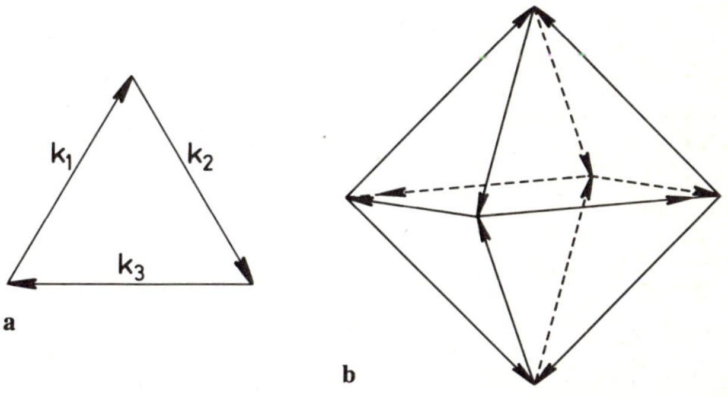


**Fig. 1 a, b.** Wave vectors of mass-density-waves in an isotropic medium. (**a**) Three equivalent vectors forming an equilateral triangle, (**b**) Six pairs of wave-vectors forming on octahedron

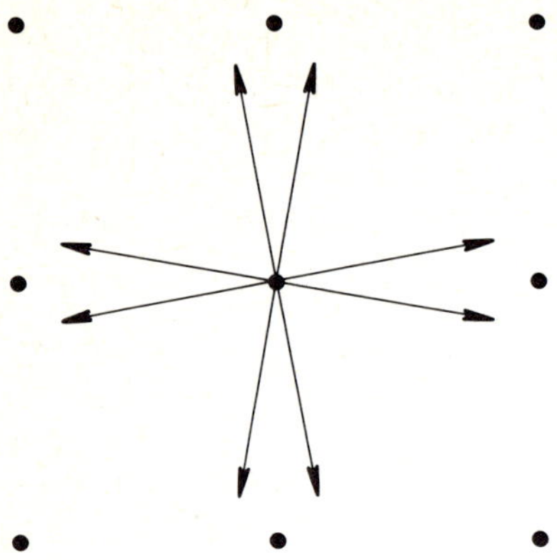

**Fig. 2.** Wave-vectors forming the "star of $\vec{q}$" for a cubic system. Note that it is not possible in general for three vectors to add up to zero or a reciprocal lattice vector

The transition will necessarily be first order, since when A becomes small enough $\varrho$ will jump to a non-zero value $\varrho_0$. The BCC structure predicted by the present argument should not be taken too seriously; higher-order terms may favor other structures. The important point is that the transition is first order because of the possibility of forming third order terms.

Furthermore, Brazovsky[3)] has pointed out that even in the absence of third order terms the transition should be discontinuous. Because of the rotational symmetry of the wave-vector, fluctuations become important near $T_c$. Eventually, fluctuations will renormalize the fourth order term of the expansion so that it becomes negative, leading to a first order transition[1)].

## 1.2 Why is the Transition not Necessarily Discontinuous for the Systems to be Considered here?

If the melting takes place on a substrate lattice in two dimensions or in a three dimensional host lattice, the rotational symmetry is broken in the liquid phase. The arguments presented above forbidding a continuous transition do not hold any more. Consider, for instance, a host lattice with a square or cubic reciprocal lattice (Fig. 2). Suppose that the wave-vector $\vec{k}_1$ describing the solid lies in an arbitrary direction in the basal plane. There cannot be an infinity of degenerate vectors, but only a finite number of vectors (eight in this case) forming the "star of $\vec{k}_1$". The coefficients $\varrho_{\vec{k}_1} \dots \varrho_{\vec{k}_n}$ of the n mass-density-waves with wave-vectors $\vec{k}_1 \dots \vec{k}_n$ constitute the n components of the Landau order parameter. Clearly, in the example shown in Fig. 2 it is not possible to add three of the eight wave-vectors to form an equilateral triangle and the Landau expansion contains no third order terms. Also, Brazovsky's fluctuation argument, resting on the full rotational symmetry, obviously is not valid when this symmetry is broken.

For the two dimensional systems to be considered here (physisorbed systems, surface reconstruction) additional complications arise. Fluctuations of the order parameter, ignored in the Landau theory, are often of crucial importance in two dimensions in determining the order of the transition, or even in determining global phase diagrams. First of all, there is a proof by Landau[4)] and Peierls[5)] that in two dimensions there cannot exist a lattice with complete long range order at any non-zero temperature, at least in the case of a smooth substrate. Kosterlitz and Thouless[6], and Nelson and Halperin[7] have argued that a phase transition, changing the nature of the correlation function, may take place anyhow.

Even if Landau theory is not valid in two dimensions, the structure of the free energy expansion can be used to *classify* the transitions[8-10]. This has to do with the concept of universality: It is believed that two transitions with the same symmetry (as expressed for example by the Landau expansion) have the same critical behavior. We shall use the Landau expansion to link together physically realizable transitions with more or less exact theories with the proper symmetry.

In three dimensions, fluctuations are important near the critical temperature. In some cases (of which we shall see a few examples) fluctuations may even change a possible continuous transition to a discontinuous one with a latent heat[11-12]. We shall see that the melting of lithium in the graphite intercalation compound $C_6Li$ belongs to the Potts universality class[13]. It has been argued that fluctuations in this case tend to make the first order transition predicted by Landau theory continuous. However, experiments indicate a weak first order melting transition[14], so it seems that in three dimensions fluctuations are not quite capable of driving the transition second order.

## 2 Melting and Solidification of Epitaxial Structures

Rare gas monolayers (He, Kr, Xe, Ar) adsorbed on graphite undergo phase transitions from high-temperature disordered liquid-like phases to low-temperature solid-like structures. We are thus dealing with the phenomenon of two-dimensional melting. The various phases and the transitions between them have been studied using neutron[15] or X-ray diffraction[16, 17], low energy electron diffraction[18] (LEED), and by measurements of vapor-pressure isotherms[19] and specific heat[20-22].

Kosterlitz and Thouless[6], and Nelson Halperin[7] have developed a theory of freely suspended two-dimensional crystals. The theory would apply directly in the case of a smooth substrate. The most important effect of the substrate is the tendency to *lock* the periodicity of the adorbate to a simple rational multiple of the substrate periodicity and form a commensurate or "registered" solid. Another effect of the substrate is to rotate the adsorbate along specific directions[23]. Although our main concern is primarily with real systems (and hence with the effects of the substrate) let us start by briefly describing the main features of the Nelson-Halperin theory. Many of the features remain when one considers melting on the substrate, provided that the period of the adsorbate is incommensurate with that of the substrate. When the system is commensurate the nature of the transition is quite different, and may be better described by means of more discrete lattice-gas models. The phase diagrams of the rare gas monolayers on graphite include melting of both commensurate and incommensurate structures.

## 2.1 Theory of Two Dimensional Melting

The Landau-Peierls theorem states that it is not possible to have a two dimensional ordered phase in a system with continuous degrees of freedom. The gapless excitations will destroy the long range order. In a two dimensional crystal, uniform translations represent such degrees of freedom. How can we then have a transition from a liquid-like to a solid-like phase in two dimensions?

Let us first consider the order-parameter correlations at low temperatures. The mass density associated with displacement $\vec{u}(\vec{r})$ of the lattice may be written

$$\varrho(\vec{r}) = A \exp i\vec{k}[\vec{r} + \vec{u}(\vec{r})] ,$$

and the density-density correlation function becomes

$$S(\vec{r}) = \langle \varrho(\vec{r})\varrho(0)\rangle = A^2 \exp i\vec{k}\cdot\vec{r}\,\langle \exp i\vec{k}\cdot(\vec{u}(\vec{r})-\vec{u}(0))\rangle \tag{2.1}$$

$$\simeq A^2 \exp i\vec{k}\cdot\vec{r}\,\exp\left\{\frac{1}{2}\langle\vec{u}(\vec{r})-\vec{u}(0)\rangle^2|\vec{k}|^2\right\}.$$

The correlation function in the exponent can be calculated in the harmonic approximation:

$$\langle\vec{u}(\vec{r}) - \vec{u}(0)\rangle^2 \sim k_B T \ln\left(\frac{r}{a}\right)$$

so the correlations decay algebraically:

$$S(\vec{r}) \sim \vec{r}^{-\eta(T)} \quad \text{with } \eta \sim T . \tag{2.2}$$

For an ordered solid $S(\vec{r})$ approaches a constant for $\vec{r} \to \infty$. For a completely disordered liquid, $S(\vec{r})$ decays exponentially for large distances. Since at low temperatures the correlation function decays with a power law there is thus a possibility of a transition from a "quasi solid" with correlation functions described by (2.1) to a completely disordered liquid, as first pointed out by Jancovici[24)].

Kosterlitz and Thouless have suggested the existence of such a transition in magnetic systems. Nelson and Halperin generalized the theory to solids. The resulting picture is: At low temperature a solid phase with power-law decay of correlation functions caused by phonons is stable. Dislocations are thermally excited, but occur in pairs only, since the free energy of forming a single dislocation is infinite. At a critical temperature, $T_m$, the small number of dislocations *dissociate,* causing the correlation functions to decay exponentially. The dissociation of the dislocations is thus responsible for destabilizing the solid.

The important point is that the transition may be *continuous* in contrast to the Landau theory. If this is the case, crystal formation is obviously quite different in two dimensions from that in three dimensions. At a second order transition there are strong precursor effect. Halperin and Nelson[7)] and Young[25)] find that as $T_m$ is approached from above the correlation length $\xi(T)$ diverges in the following way:

$$\xi(T - T_m) \sim \exp(C/(T - T_m)^\nu) \tag{2.3}$$

with

$\nu = 0.36963\ldots$ .

Now, even at temperatures slightly above $T_m$, there is still some ordering left. Let $\Theta(\vec{r})$ denote the local rotation at the position $\vec{r}$. It turns out that the correlation function for the rotational order parameter $\langle \exp 6i\,\Theta(\vec{r})\rangle$ decays algebraically:

$$\langle \exp 6i(\Theta(\vec{r}) - \Theta(0))\rangle \sim \vec{r}^{-\eta_{rot}} \tag{2.4}$$

The factor 6 arises because a crystal rotated $\frac{2\pi}{6}$ is indistinguishable from the non-rotated crystal. To completely destroy the *rotational* "quasi order" another transition is needed. Nelson and Halperin show that such a transition can be brought about by dissociation of *disclinations*. Above the disclination-unbinding transition both the density correlations decay exponentially.

One cannot say that the theory has been confirmed by independent calculations. In fact, molecular dynamics calculations indicate a first-order transition for Lennard-Jones potentials[26)], and no sign of a second transition. The discontinuity of the energy at the transition, however, is quite small compated to that of a traditional melting process.

## 2.2 Melting and Solidification of Incommensurate Physisorbed Systems

Figure 3 shows commensurate and incommensurate ordered solids similar to those found in the rare-gas monolayers on graphite. Figure 3a shows the so-called "$\sqrt{3}$ structure" where 1/3 of the graphite hexagons are occupied by a rare gas atom. This may represent the ordered phase of krypton at certain densities. The Xenon atom is slightly too large to form the $\sqrt{3}$ structure, so the ordered phase is an *incommensurate* solid (Fig. 3b). Similarly, the argon atom is slightly too small to form the $\sqrt{3}$ structure, and the incommensurate phase with a slightly higher density of atoms is favorable.

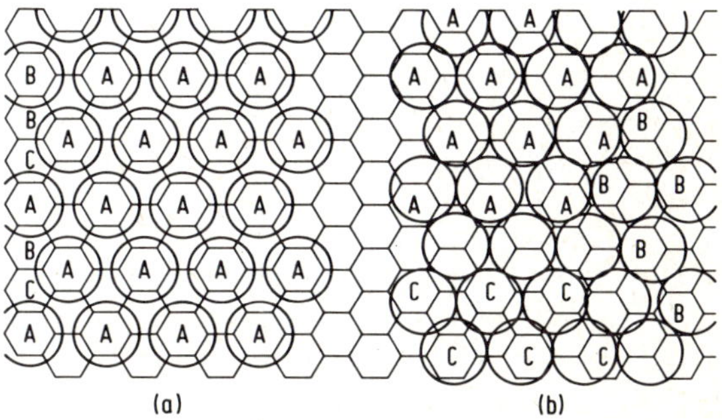


**Fig. 3a, b.** Rare gas monolayers adsorbed on graphite. (**a**) Commensurate "$\sqrt{3}$ structure" (**b**) Incommensurate structure. The honeycomb lattice represents the graphite lattice and the circles the rare gas atoms

An incommensurate lattice may be shifted relative to the stubstrate without crossing an energy barrier[27]. There is thus a continuous symmetry and the Landau-Peierls argument applies.

What is the effect of the periodic potential from the graphite substrate on the Nelson-Halperin theory?

The substrate tends to align the crystal along some preferred direction. The substrate provides a field which is conjugate to the rotational order parameter $\langle \exp 6i\,\Theta \rangle$, and there will be some long rage rotational order at any temperature. There is thus no need for the *second* transition to destroy rotational correlations. In a sense the situation is very similar to that of a ferromagnet in a field, where there is some long range order (magnetization) even at the highest temperatures, and there is no phase transition in the thermodynamic sense.

There exist two synchrotron X-ray experiments by Heiney et al.[16], and by Nielsen et al.[28] on Xe and Ar adsorbed on graphite. The phase diagram is shown schematically in Fig. 4a. It consists of ordered regimes, disordered regimes, and regimes with coexistence of solid and liquid phases. The transition in question here is the one indicated by a broken line. At a first order transition, phase separation would take place in a small regime between the solid and liquid phases. For Xe both groups find indications of a continuous transition, at least for part of the transition line. Nielsen et al. also find a second-order-like transition for argon. Heiney et al.[16] find diffraction peaks characteristic of power-law correlation functions for temperatures below $T_m$, and Lorenzian line shapes characteristic of exponential decay at temperatures above $T_c$. Figure 4b shows diffraction peaks for temperatures slightly above and below $T_m \simeq 140$ K. Again, the most interesting feature is that there are no discontinuities or hysteresis near $T_m$.

In addition to these experiments on Xe and Ar on graphite there is also a neutron diffraction experiment on methane ($CH_4$) adsorbed on graphite (by Dutta et al.[29]) indicating a continuous melting transition. However, the experimental resolution obtained by the synchrotron measurements is far superior to that obtained by the neutron scattering measurements.

In conclusion, there is some experimental evidence that melting of incommensurate crystals may take place through a transition which is at least almost continuous. It is clear from both experiments and calculations that in two dimensions fluctuation have a strong tendency towards suppressing the first order transition predicted by Landau theory.

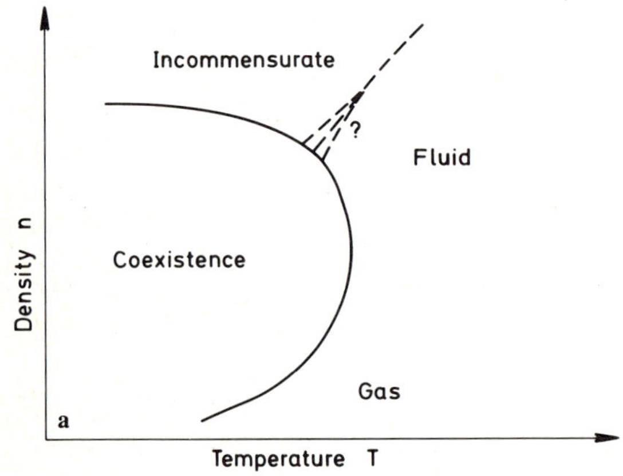


**Fig. 4. a** Schematic phase diagram for xenon or argon adsorbed on graphite. The melting transition of interest here is along the *broken line*

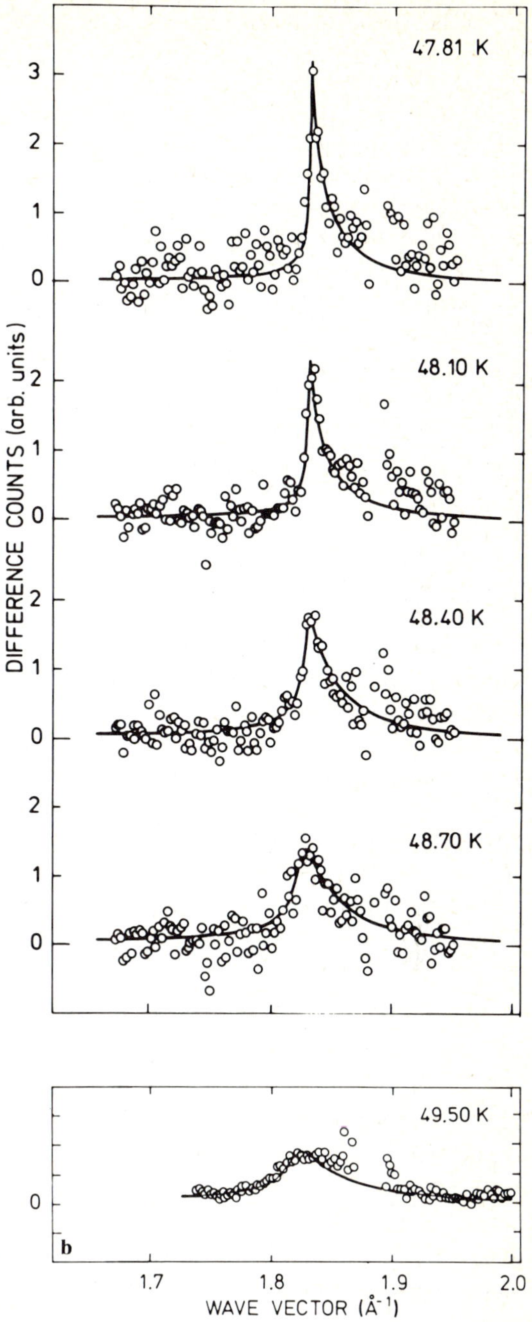


**Fig. 4. b** Scattering profiles from Ar monolayers on graphite measured through the melting transition (Nielsen et al., Ref. 28)

## 2.3 Melting of Commensurate Epitaxial Systems: Krypton on Graphite

For commensurate ordered structures there is no continuous symmetry and there is a possibility of having complete long rage order.

For such structures, the Landau expansion of the free energy can be used to classify the melting transitions. The philosophy is as follows: On the basis of experimental information on the high and low-temperature phases the Landau free energy functional is constructed. The structure of the expansion determines the universality class of the transition. The critical behavior to be expected can then be determined by comparing with theories having the same symmetry.

Alexander[8] was the first to apply this philosophy to a specific system, namely He adsorbed on graphite. The He atoms order in the same structure as krypton on graphite, the "$\sqrt{3}$ structure" shown in Fig. 3. The reciprocal lattice vector (the wave-vector of the ordering mass-density wave) is

$$\vec{q} = \frac{2\pi}{a}(1/\sqrt{3}, 0) \tag{2.5}$$

where a is the graphite lattice constant. The "star" of this wave vector consists of only two vectors, $\pm\vec{q}$. The vectors related to $\vec{q}$ through the threefold symmetry are identical to $\vec{q}$ or $-\vec{q}$ apart from a reciprocal lattice vector. The dimensionality of the order-parameter is thus $n = 2$, and we are dealing with an XY-like system. The two components, $\psi_1$ and $\psi_2$ describe mass density waves

$$\begin{aligned} \varrho(r) &= \psi_1 \cos\vec{q}\cdot\vec{r} \\ \varrho(r) &= \psi_2 \sin\vec{q}\cdot\vec{r} \end{aligned} \tag{2.6}$$

respectively.

The most general Landau expansion having the symmetry of the graphite lattice is

$$F = \frac{1}{2} r(\psi_1^2 + \psi_2^2) + v(\psi_1^3 - 3\psi_1\psi_2) + \dots . \tag{2.7}$$

There is a discrete lattice model, namely the three state Potts model which has the same symmetry[8], so the melting transition belongs to the "universality class of the three state Potts model". For $v < 0$ the ordered state is characterized by the densities

$$\begin{aligned} \varrho_1 &= A\cos\vec{q}\cdot\vec{r} \\ \varrho_2 &= -\frac{1}{2}(A\cos\vec{q}\cdot\vec{r} + \sqrt{3}\,A\sin\vec{q}\cdot\vec{r}) \\ \varrho_3 &= -\frac{1}{2}(A\cos\vec{q}\cdot\vec{r} - \sqrt{3}\,A\sin\vec{q}\cdot\vec{r}) . \end{aligned} \tag{2.8}$$

These densities represent the three equivalent positions (A, B, C in Fig. 3) for the krypton lattice, or the three equivalent states in the Potts model. The phase transition of the Pott's model in known to be *second order*[30)] despite the third order terms. Moreover, there is a model belonging to the same universality class, the hard hexagon model, which has been solved by Baxter[31)]. The exponents $\beta$ and $\alpha$, characterizing the order parameter, A, and the specific heat c near $T_c$,

$$A \sim (T_c - T)^{\beta}$$
$$c \sim (T - T_c)^{\alpha}$$

were found to be $\beta = 1/9$ and $\alpha = 1/3$ .

Figure 5 shows the phase diagram of krypton adsorbed on graphite (Birgeneau et al., Ref. 16). Most of the information comes from specific heat studies by Butler et al.[22)]. Birgeneau et al.[16)] found that for coverages near one monolayer the transition from C to fluid seemed indeed to be second order. However, because of finite size smearing effects it is difficult to extract information on the critical indices. To understand the global phase diagram it is of course not sufficient to use the universality arguments above. Berker, Ostlund and Putnam[32)] have studied a microscopic Potts lattice gas model using Migdal renormalization group techniques, and obtained good agreement with experiments for sub-monolayer coverages. In their model the krypton atoms were confined to be exactly at the graphite hexagons (the gas was confined to be on a lattice). Among other things they found that for small coverages the transition is first order, in agreement with experiment. At higher coverages where an incommensurate squeezed phase occur, the lattice gas model is insufficient.

The Potts transition is extremely interesting since it is very well-documented, both theoretically and experimentally, that the melting transition is continuous with not latent heat. The solidification process is reversible, so the formation of the 2d crystal is com-

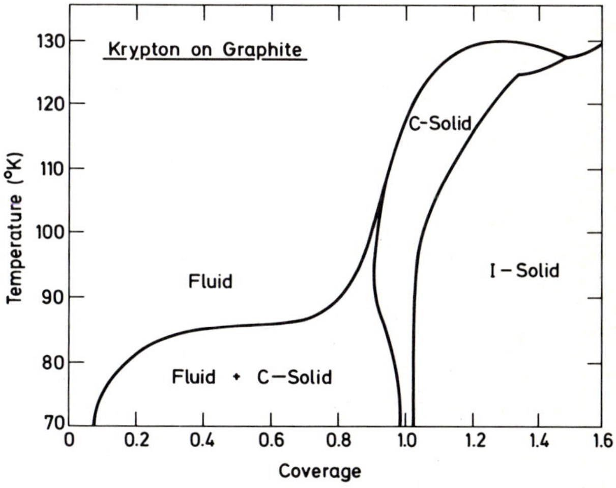


**Fig. 5.** Phase diagram for krypton adsorbed on graphite (Birgeneau et al. Ref. 16. The phase boundaries come primarily from the specific heat studies of Butler et al., Ref. 22)

pletely different from the non-quilibrium crystal-growth process usually encountered in three dimensions.

Domany et al.[9] have classified a number of possible two-dimensional order-disorder transitions involving low-order commensurate phases. They find transitions belonging to a number of universality classes including two- and three-state Pott's models, Ising models, and XY models with cubic symmetry.

Higher order commensurate phases are expected to melt in a different way. Halperin and Nelson[7], following ideas of José et al.[33] find that there is an intermediate "floating" phase between the low-temperature completely ordered phase and the high-temperature disordered liquid phase. The floating phase has power-law decay of the order parameter, in a way very similar to the incommensurate phase.

As a parameter (such as pressure or density) is varied the system may pass through several commensurate phases. The structure of the global phase diagram could look like the diagram shown in Fig. 6. Note that the fluid phase may penetrate all the way down to T = 0 near the "$\sqrt{3}$" phase. The incommensurate phase near the $\sqrt{3}$ phase may be described as a commensurate phase with a network of compressed walls[34]. Coppersmith et al.[35], extending an argument by Villain and Bak[36], argued that the network will melt before the transition to the commensurate phase take place.

## 2.4 Surface Reconstruction

Low energy electron diffraction (LEED) studies[37–39] have shown that surfaces of several metals are recontructed, i.e. surface atoms are arranged in a different way from those in the bulk material. When the temperature is raised the surface may undergo a transition from the recontructed state to the ideal 1 × 1 state. In some cases these transitions can be thought of as a melting or solidifcation of the upper layer of atoms. The transition can then be described in a way similar to that of the physisorbed systems above[10, 40].

For instance, the surface layer of atoms on the Ir, Pt, and Au [110] phases seems to undergo a transition from the high temperature disordered phase to a solid phase where

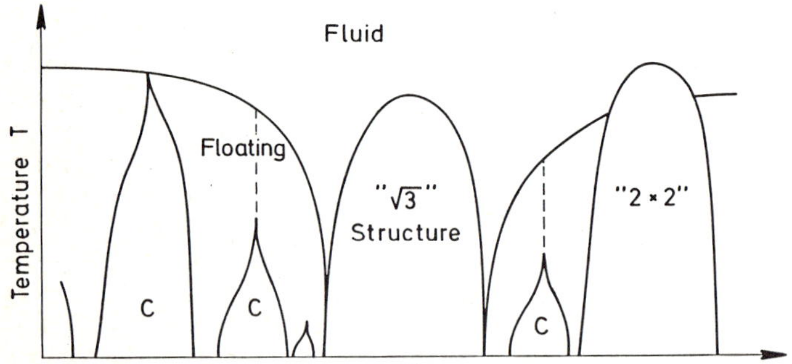


**Fig. 6.** Schematic global phase diagram of a two-dimensional epitaxial system. C: commensurate phase. Note that a floating phase generally separates the high-order commensurate phases from the fluid phase

every second [110]-row of atoms is missing. The ground state is two times degenerate and the surface melting transition turns out to be an Ising transition. The two-dimensional Ising model has been solved and the critical indices of the continuous transition are known[41]. In principle, the properties of the transition could be studied by LEED or Helium scattering, but no such measurements have been performed yet. The reconstruction transitions on the [100]-surfaces of W[39], or Mo[37] are described by more complicated XY-like models[40], and again it would be interesting to test predictions of the critical behaviour experimentally.

# 3 Melting and Solidification in Three Dimensional Intergrowth Compounds

There exist a number of three dimensional systems where one crystal-lattice solidifies inside another. This is, in some sense, the 3d analog of the 2d monolayers on graphite. It has been suggested to denote such crystals "intergrowth" compounds[42]. Here we shall consider two systems, one quasi-two-dimensional (graphite intercalation compounds) and one quasi-one-dimensional (mercury chain compounds). It will be seen that melting transitions with a number of unusual properties take place.

## 3.1 Melting in Graphite Intercalation Compounds

Structurally, the intercalation compounds consist of hexagonal graphite honeycomb layers between which there are layers of, for instance, alkali metal atoms. Figure 7 shows a "stage two" compound where the metal layers are separated by two graphite layers. At high temperature there is no long range order in the metal system, but at low tempera-

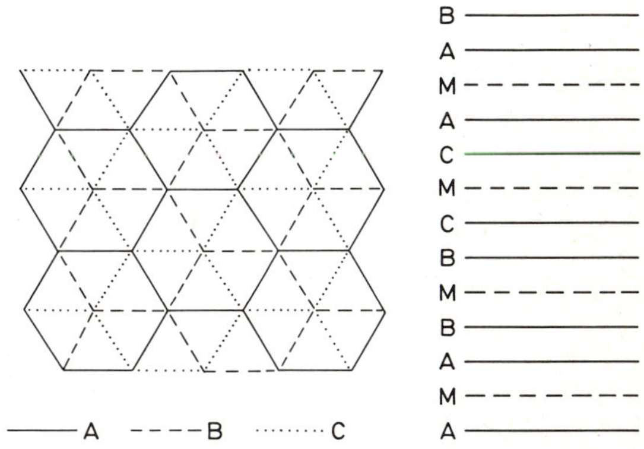


**Fig. 7.** Stacking of metal and graphite layers in intercalation compounds (here stage 2). The left-hand side shows the three equivalent positions of the graphite layers. Within layers, the intercalated atoms may order for instance like the structures shown in Fig. 3

tures there is often a transition into a structure where the metal ions form a regular lattice which may be commensurate or incommensurate with the graphite lattice. Within the layers the ordered structures are very similar to those shown in Fig. 3. The melting transitions in intercalation compounds have been analyzed by Bak and Domany[13, 43–44].

As emphasized in the introduction, the degeneracy of the wave-vector of the order-parameter is lifted because of the existence of the graphite lattice. In the stage-1 compounds the graphite layers are stacked in a uniform way AAAA. In $LiC_6$ the Li atoms order in a structure where the Li atoms occupy one of the three symmetric sets of lattices, $\alpha$, $\beta$ or $\gamma$ in the "$\sqrt{3}$ structure"[45]. The stacking sequence is simply $\alpha\alpha\alpha\cdots$ ($\beta\beta\beta\cdots$, or $\gamma\gamma\gamma\cdots$). The wave-vectors of the Li-structure are

$$\pm\vec{q} = \pm\frac{2\pi}{a}(1/\sqrt{3},0,0) \tag{3.1}$$

Note the similarity with (2.5). The order-parameter has two components, and the three ordered states are given by (2.8), as for the ordering of krypton on graphite. Of course, the mass density waves are now three-dimensional.

The Landau expansion takes exactly the same form as for krypton on graphite, and we can immediately identify the transition as a 3d-three-state Pott's transition. Despite the fact that Landau theory predicts a first-order transition because of the third-order term this transition has been the matter of controversy and conflicting theories[46]. Recently, some consensus has been reached that the transition is discontinous, based on Monte-Carlo analysis[47] and experiments on very different physical systems exhibiting Potts transitions[48].

Very recently the transition has been studied by means of calorimetry and neutron-diffraction by Robinson and Salamon[49]. They find a first-order transition. Figure 8 shows the development of the order parameter as a function of temperature. The Monte-Carlo results of Knak-Jensen and Mouritsen[47] are also shown. The curves clearly show a discontinuity of the order parameter. The transition has some unusual features. Both

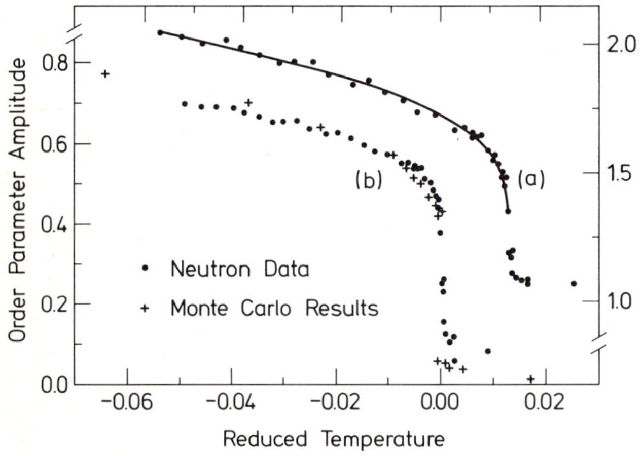


**Fig. 8.** Neutron diffraction measurements of the order parameter in $C_6Li$ (Robinson and Salamon, Ref. 49). Monte Carlo results by Knak-Jensen and Mouritsen are also shown (Ref. 47)

specific heat and the order-parameter correlation exhibit strong precursor effects. which seem to be characteristic for the Pott's transition. The transition is "almost" continuous.

The intralayer ordering of the stage 1 compounds $C_6Eu$, $C_6Yb$, and $C_6Ba$ is the same as for $C_6Li$. However, the stacking sequence is $\alpha\beta\alpha\beta\cdots$ and not $\alpha\alpha\alpha\alpha\cdots$ [50]. The wave-vectors describing the ordered structure are

$$\pm\vec{q} = \pm\left(1/\sqrt{3},\ 0,\frac{1}{2}\right). \tag{3.2}$$

Because of the component in the c-direction it is not possible to add up three $\vec{q}$ vectors to form a reciprocal lattice vector and the Landau expansion has no third order invariants:

$$F = \frac{1}{2}\, r\,(\psi_1^2 + \psi_2^2) + u\,(\psi_1^2 + \psi_2^2)^2 . \tag{3.3}$$

This is the expansion of the 3d-XY model. We may therefore expect a continuous melting transition characterized by the exponents of the XY-model:

$$\psi \sim (T_c - T)^{\beta} \tag{3.4}$$

with $\beta \approx 0.33$ [51].

It would be interesting to study the transition experimentally since it seems that there is a good possibility of observing a well-defined continous melting transition for the first time in a three-dimensional system.

The stage-one compounds $C_8Rb$ and $C_8C_s$ also order in commensurate structures[52–53]. The analysis of Bak and Domany[43] shows that the transition in $C_8Cs$ belongs to the universality class of a complicated 6-component n-vector model with third order terms. The transition in $C_8Rb$ has the symmetry of the Heiseberg model with cubic symmetry. The transition is expected to be discontinuous because of the sign of the anisotropy.

The stage-two compounds $C_{24}Cs$, $C_{24}Rb$ and $C_{24}K$ order in incommensurate structures[54]. For the ideal systems, the order-parameters are complicated high-dimensional ones[43]. Renormalization group arguments[11] indicate that fluctuations drive the transition first order despite the fact that Landau theory permits continuous transitions. Indeed, experiments[54] indicate that correlation lengths are increasing as the transition is approached from above. However, the graphite lattice is not stacked in a regular way in these systems, and it is expected that stacking faults will tend to smear the transition and make it appear more continuous. Also, the stacking faults will destroy the long range order below $T_c$ to some extent, and make the ordered phase more "glassy"[43].

Figure 9 shows the "in-plane" correlation length for $C_{24}Cs$ as a function of temperature as measured by Clarke. At no point does the correlation length diverge. The interlayer correlations increase rapidly around 170 K.

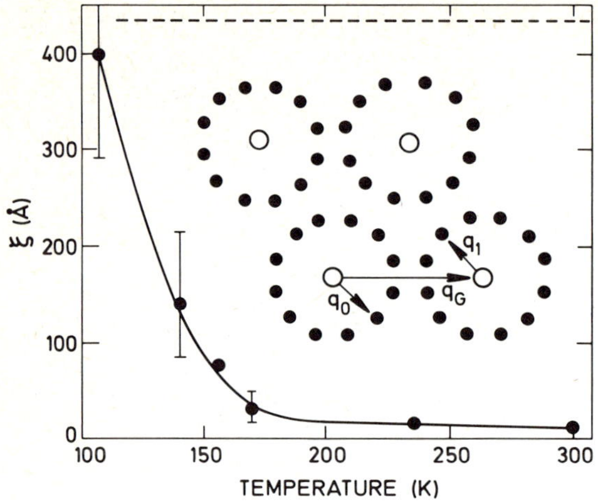


**Fig. 9.** Temperature dependence of the correlation length in the stage-2 compound $C_{24}Cs$ (Clarke, Ref. 54)

## 3.2 Melting of Mercury Chains in $Hg_{3-\delta}AsF_6$

The structure of the mercury chain compound $Hg_{3-\delta}AsF_6$ is formed by a body-centered tetragonal lattice of $AsF_6$ through which pass two non-intersecting arrays of mercury ions parallel to the basal plane edges of the $AsF_6$ host lattice (Fig. 10). At high temperature the mercury atoms form an "incommensurate" liquid with no long-range atomic positional order, but at $T_c$ = K there is a phase transition into a structure where the mercury chains form an incommensurate ordered lattice (Fig. 11)[55].

Note that in the ordered phase the mercury atoms are confined to (110) planes, so the solidification can be described as the formation of a mass-density wave with wave-vector $q_1 = (\delta, \delta, 0)$. The star of this vector has four components

$$q_1 = \pm(\delta, \delta, 0)$$

and (3.5)

$$q_2 = \pm(\delta, -\delta, 0) ,$$

so the order-parameter has n = 4 components[56], $\psi_{\pm 1}$ and $\psi_{\pm 2}$. The Landau expansion takes the form:

$$\begin{aligned} F &= \frac{1}{2} r(\psi_1^2 + \psi_{-1}^2 + \psi_2^2 + \psi_{-2}^2) \\ &\quad + u(\psi_1^2 + \psi_{-1}^2 + \psi_2^2 + \psi_{-2}^2)^2 \\ &\quad + v(\psi_1^2 + \psi_{-1}^2)^2(\psi_2^2 + \psi_{-2}^2) . \end{aligned} \tag{3.6}$$

As can be seen there are no third order term so a continuous transition seems possible. The transitions from a paramagnetic state to a helical ordered structure in Ho, Dy and Tb

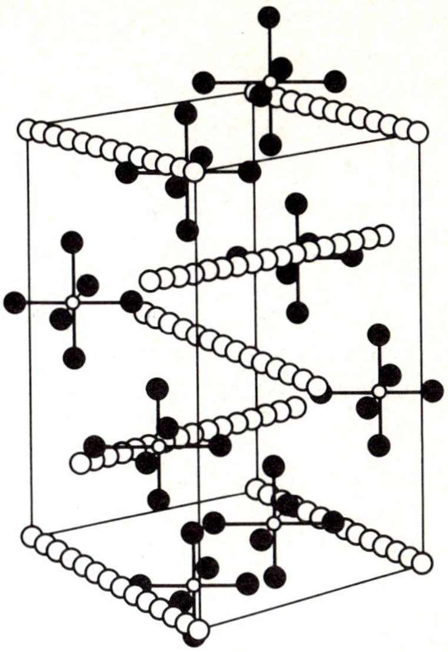

**Fig. 10.** Crystal structure of the mercury chain compound $Hg_{3-\delta}AsF_6$. The Hg ions are shown schematically

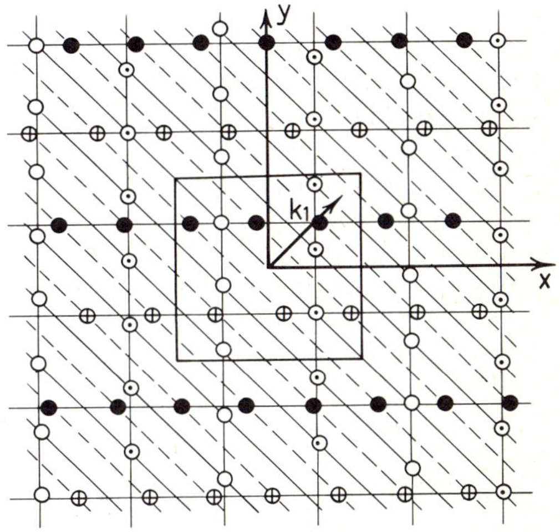


**Fig. 11.** Ordered structure of $Hg_{3-\delta}AsF_6$. The circles indicate the projections of Hg-atoms on the basal plane. The positions along the tetragonal axis are as follows. ○: z = 0; ●: z = 1/4, ⊙: z = 1/2, ⊕: z = 3/4

are described by the same "Hamiltonian". Depending on the sign of the anisotropy term v, F may favour *either* a structure formed by a single mass-density wave, or a structure which is formed as the superposition of two-orthogonal waves (the double-$\vec{q}$ structure). A renormalization group analysis shows that only in the latter case is a continuous transition possible[56)]. The experiments indicate that the single mass-density-wave structure in Fig. 2 is indeed the stable one, so a fluctuation-induced first-order transition is expected. Indeed, experiments indicate hysteresis around the transition temperature[55)], but apart from that the transition is almost continuous. For instance, correlation-lengths grow rapidly as the transition is approached, as is usual for fluctuation-induced first-order transitions. Also, below $T_c$ the intensities of the Bragg peaks seem to go continuously to zero at $T_c$, seemingly with mean -field exponents.

The three-dimensional intergrowth compounds exhibit a variety of unusual melting transitions. As we have seen, some of these are expected to belong to universality classes of well-defined and well-studied models in statistical mechanics. Some of the systems may exhibit continuous melting, and even in the cases where melting is discontinuous there may be strong precursor effects, since in some cases the discontinuity is caused by fluctuations. For other systems fluctuations may almost drive the transition continuous (the Potts systems). The processes by which solidification takes place are thus very different from the nucleation and crystal growth processes usually encountered in three dimensional systems. I expect that the detailed study of melting and solidification in intergrowth compounds will remain a very active field of research, since the detailed properties near melting are known only in very few cases.

## 4 References


1. Landau, L. D., Lifshitz, E. M.: Statistical Physics, Pergamon, New York 1968
2. Alexander, S., McTague, J. P.: Phys. Rev. Lett. *41,* 702 (1978)
3. Brazovsky, S. A.: ZhETF *68,* 42 (1975) [Sov. Phys. JETP *41,* 85 (1975)]
4. Landau, L. D.: Phys. Z. Sowjetunion II, 26 (1937)
5. Peierls, R. E.: Ann. Inst. Henri Poincaré 5, 177 (1935)
6. Kosterlitz, J. M., Thouless, D. J.: J. Phys. C *6,* 1181 (1973)
7. Nelson, D. R., Halperin, B. I.: Phys. Rev. B *19*, 2457 (1979)
8. Alexander, S.: Phys. Lett. A *54*, 353 (1975)
9. Domany, E., Schick, M., Walker, J. S., Griffiths, R. B.: Phys. Rev. B *18*, 2209 (1978)
10. Bak, P.: Physica *99 B*, 325 (1980)
11. Bak, P., Mukamel, D., Krinsky, S.: Phys. Rev. Lett. *36,* 52 (1976)
12. Mouritzen, O. G., Knak-Jensen, S. J., Bak, P.: ibid. *39*, 629 (1977)
13. Bak, P., Domany, E.: Phys. Rev. B *20*, 2818 (1979)
14. Robinson, D. S., Salamon, M. B.: Phys. Rev. Lett. *48,* 156 (1982)
15. See, for instance Kjems, J. K., Passell, L., Taub, H., Dash, J. G., Novaco, A. D.: Phys. Rev. B *13*, 1446 (1976); McTague, J. P., Nielsen M., Passell, L.: Ordering in strongly fluctuating systems, p. 195 (ed. Riste, T.) Plenum, Yew York 1980
16. Stephens, P. W., Heiney, P., Birgenau, R. J., Horn, P. M.: Phys. Rev. Lett. *43,* 47 (1979); Birgeneau, R. J., Hammons, E. M., Heiney, P., Stephens, P. W.; Ordering in two dimensions (ed. Sinha, S.) p. 29. North-Holland, New York 1980; Heiney, P. A., Birgeneau, R. J., Brown, G. S., Moncton, D. E., Stephens, P. W.: Phys. Rev. Lett. *48*, 104 (1982)
17. Nielsen, M., Als-Nielsen, J., Bohr, J., McTague, J. P.: Phys. Rev. Lett. *47*, 582 (1981)
18. Chinn, M. D., Fain, S. C. Jr.: ibid. *39,* 146 (1977); Fain, S. C. Jr., Chinn, M. D., Diehl, R. D.: Phys. Rev. B *21,* 4170 (1980)
19. Thomy, A., Duval, X.: J. Chim. Phys. *66,* 1966 (1969)
20. Bretz, M., Dash, J. G., Hickernell, D. C., McLean, E. O., Vilches, O. E.: Phys. Rev. A *8,* 1589 (1973)
21. Bretz, M.: Phys. Rev. Lett. *38,* 501 (1977)
22. Butler, D. M. Litzinger, J. A., Stewart, G. A.: ibid. *44,* 466 (1980)
23. Novaco, A. A., McTague, J. P.: ibid. *38*, 1286 (1977)
24. Janocovici, B.: ibid. *19,* 20 (1967)
25. Young, P.: Phys. Rev. B *19,* 1855 (1979)
26. Toxvaerd, S.: Phys. Rev. Lett. *44,* 1002 (1980)

27. For a review on incommensurate systems. See, for instance, Bak, P.: Rep. Progress Phys. *45*, 587 (1982)
28. Nielsen, M., Als-Nielsen, J., Bohr, J., McTague, J. P.: Phys. Rev. Lett. *41*, 582 (1981)
29. Dutta, P., Sinha, S. K., Vora, P., Nielsen, M., Passell, L., Bretz, M.: Ordering in two dimensions (ed. Sinha, S.) p. 169. North-Holland, New York 1980
30. Baxter, R. J.: J. Phys. C *6,* L445 (1973)
31. Baxter, R. J.: J. Phys. A. *13,* L61 (1980)
32. Berker, A. N., Ostlund, S., Putnam, F. A.: Phys. Rev. B *17*, 3650 (1978)
33. José, J., Kadanoff, L. P., Kirkpatrick, S., Nelson, D. R.: ibid. *16,* 1217 (1977)
34. Bak, P., Mukamel, D., Villain, J., Wentowska, K.: ibid. *19*, 1610 (1979)
35. Coppersmith, S. N., Fisher, D. S., Halperin, B. I., Lee, P. A., Brinkman, W. F.: Phys. Rev. Lett. *46,* 549 (1981)
36. Villain, J., Bak, P.: J. Physique *42,* 657 (1981)
37. Estrup, P. J., Barker, R. A.: "Ordering in two dimensions" p. 39 (ed. Sinha, S.) North-Holland, New York 1980
38. Blakeley, D. W., Somerjai, G. A.: Surf. Sci. *65,* 419 (1977)
39. Debe, M. K., King, D.: Surf. Sci. *81*, 193 (1979)
40. Bak, P.: Solid State Commun. *32,* 581 (1979)
41. Onsager, L.: Phys. Rev. *65,* 117 (1945)
42. Axe, J. D., Bak, P.: Phys. Rev. B *26*, 4963 (1982)
43. Bak, P., Domany, E.: ibid. *23,* 1320 (1981)
44. Bak, P.: Phys. Rev. Lett. *44,* 889 (1980)
45. Guerard, G., Herold, A.: Carbon *13,* 337 (1975)
46. Ditzian, R. V., Oitmaa, J.: J. Phys. A *7,* L61 (1974); Straley, J. P.: J. Phys. A *7*, 2173 (1974); Golner, G.: Phys. Rev. B *8*, 3419 (1979)
47. Knak-Jensen, S. J., Mouritsen, O. G.: Phys. Rev. Lett. *43,* 1736 (1979)
48. Barbara, B., Rossignol, M. F., Bak, P.: J. Phys. C *11*,, L183 (1978); Aharony, A., Müller, K. A., Berlinger, W. B.: Phys. Rev. Lett. *38*, 33 (1977)
49. Robinson, D. S., Salamon, M. B.: ibid. *48*, 156 (1982)
50. El Makrini, M., Guerard, D., Lagrange, P., Herold, A.: Physica B + C *99,* 481 (1980)
51. Fisher, M. E.: Rev. Mod. Phys. *46,* 597 (1974)
52. Ellenson, W. D., Semmingsen, D., Guerard, D., Onn, D. G., Fischer, J. E.: Mater. Sci. Eng. *31,* 137 (1977)
53. Clarke, R., Caswell, N., Solin, S.: Phys. Rev. Lett. *42*, 61 (1979)
54. Clarke, R.: Ordering in two dimensions. (ed. Sinha, S.) p. 53, North-Holland, New York 1980; Zabel, H.: Idem, p. 61; Suzuki, M., Ikeda, H., Suematsu, H., Endoh, Y., Shiba, H., Hutchings, M. T.: J. Phys. Soc. Japan *49,* 671 (1980)
55. Hastings, J. M., Pouget, J. P., Shirane, G., Heeger, A. J., Miro, A. D., McDiarmid, A. G.: Phys. Rev. Lett. *39,* 1484 (1977); Pouget, J. P., Shirane, G., Hastings, J. M., Heeger, A. J., McDiarmid, A. G.: Phys. Rev. B *18*, 3645 (1978)
56. Bak, P.: Phys. Rev. B *20*, 2209 (1979)

# Microscopic Theory of the Growth of Two-Component Crystals

**Werner Haubenreisser and Hubert Pfeiffer**

Zentralinstitut für Festkörperphysik und Werkstofforschung der Akademie der Wissenschaften der DDR, Institutsteil für magnetische Werkstoffe, Helmholtzweg 4, DDR-6900 Jena

*A unified interface theory of crystal growth is formulated on the basis of a master-equation approach to obtain kinetic or rate equations for the description of growing faces of two-component crystals in terms of a Kossel (solid – on – solid) lattice model of the fluid-solid interface. The kinetics of binary Kossel crystal growth is introduced as a stochastic process taking into account fluid ⇄ solid phase transitions, diffusion processes in the fluid and solid phases as well as short- and long-range order effects. Equilibrium and nonequilibrium properties are described in different statistical approximation schemes (single-particle, pair and Bragg-Williams (BW) approximation). Results of pair approximation agree satisfactorily with a Monte Carlo computer simulation. Also special growth features of binary Kossel crystals can be understood within the frame of the BW approximation. A further progress in the application of a lattice-based description to problems of real crystal growth requires a more consistent incorporation of experimental data. But a deeper microscopic understanding of the growth features and peculiarities of one- and multi-component systems can only be expected within the frame of a molecular theory of the solid-fluid interface.*

# 1 Introduction

The growth of crystals containing two or more components plays an important role in many modern industrial applications[1]. The optimal control of these technologies is closely connected with an understanding of microscopic processes during crystallization. The first basic contributions towards a microscopic modelling of the growth characteristics and growth peculiarities of multicomponent systems were given in the pioneering works by Chernov[2,3], Chernov and Lewis[4] and Temkin[5,6] where mainly the growth of a binary (multicomponent) chain and the growth due to the kink movement in one-, two- and three-dimensional binary systems were treated. Further considerable progress has been made recently in the development of microscopic interface theories of the growth of multicomponent systems mainly through the works of Cherepanova and coworkers[7] to[18], Temkin[19] to[21], Pfeiffer and Haubenreisser[22,23], and Saito and Müller-Krumbhaar[24,25] where equilibrium and nonequilibrium properties in the growth of crystal surfaces were studied. Especially, models describing rough interfaces in the atomic scale have been applied to the study of two-component crystal growth (see[18,21,23]).

One of the main aims of this article is to attempt to represent a microscopic two-component crystal growth theory within a unified description where an analytical treatment is aspired in order to discuss the most frequently used approaches and statistical approximation schemes. Therefore, we deal with basic features and basic approaches of a unified microscopic theory of the growth of an inhomogeneous multicomponent system in the framework of a multilayered fluid-solid lattice interface model under solid-on-solid (SOS) restriction and discuss its applications and limitations.

Using a lattice-based description of the fluid-solid interface (as well as of the structural properties of the fluid (F) and solid (S) bulk phases), a two-phase multicomponent system is characterized by a given common rigid lattice structure where any lattice site, whose symmetry coincides with that of a given lattice inside the crystalline solid bulk phase, corresponds to a unit cell which can be occupied either by a solid $S\alpha$ or a fluid particle $F\alpha$ of each chemical component $\alpha\,(= A, B, \ldots)$ of the system. The fluid-solid interface is here considered to be that transition region separating two coexisting SOS-bulk phases S and F in which the properties differ from those of the adjoining states $m = F, S$ of particles $m\alpha$ within the interface. SOS restriction means precisely that isolated solid clusters in the bulk fluid and fluid inclusions within the bulk solid as well as overhanging interface configurations are excluded. For computational convenience, only a simple cubic (s.c.) lattice structure is usually considered. In this case, the SOS-lattice model resembles the famous Kossel crystal, a familiar model used in crystal growth theories (see[26–31]). Furthermore, the fluid-solid interface is subdivided into monoatomic layers parallel to a $\{100\}$ plane of the growing crystal face which is perpendicular to the growth direction $\langle 100 \rangle$. Such a multilayered SOS-lattice fluid-solid interface model is the starting point of a microscopic multicomponent growth theory in which the growth of perfect surfaces can be described[7] to[23] (in order to include surface dislocations, steps or other topological details, other growth models must be used (see e.g.[26,31–34])). Moreover, it should be noted that in this growth theory only the interface kinetics, i.e. the interface processes, are included whereas the remaining bulk kinetics, i.e. the mass and heat transport in the pure fluid and solid bulk phases, must be taken into account additionally (see e.g.[1,13,18,21,35,36]). The interface kinetics can formally be separated into a relaxation-type kinetics due to first-order (fluid $\rightleftarrows$ solid) phase transitions and a diffusion-type kinetics due to particle exchange processes within and/or between the fluid and solid states of the interface. In[7–23] the first-order phase transitions occurring during crystallization are modelled by single-site one-particle attachment and detachment processes of fluid $F\alpha$ and solid $S\alpha$ particles within the SOS-lattice interface and described in the concept of the order parameters (see, e.g.[37, 41]). The order parameter (or a set of such parameters) gives the orders of the relevant degrees of freedom of the system in question and also describes the degree of order of the system on a macroscopic scale; it is thus a macroscopic variable. Such an order parameter concept is suitable for the description of spatial correlations, i.e. short-range order (SRO) and/or long-range order (LRO) of a multicomponent system during crystallization. To introduce the order parameter concept into a microscopic theory of multicomponent growth we consider an equation for the probability distribution function of the occupation numbers of fluid and solid particles of type $F\alpha$, $S\alpha$ allowing also for fluctuations ("stochastic equations"). The particle numbers are random variables, and we can obtain from these stochastic equations (master equations) via suitable averaging equations for the expectation (averaged) values of occupation numbers and of higher moments (correlations, fluctuations) in the occupation numbers of particles ("kinetic or rate equations"). As order parameter a space- and time-dependent macroscopic (averaged) concentration variable is used for the description of the SRO and LRO of particles $m\alpha$ ($m = F, S$; $\alpha = A, B$) within the SOS-lattice fluid-solid interface. At present, the starting point of a unified microscopic interface theory of one- and multicomponent Kossel crystal growth is a master-equation treatment in the framework of a multilayered SOS-lattice fluid-solid interface model using the concept of order parameters. Of course, a complete quantitative treatment of the interface kinetics is only possible by Monte Carlo computer simulations[16, 18, 38, 39] whereas the analytically obtained results are determined by the used approaches and approximation schemes[18, 21, 23]. This means that the resulting kinetic or rate equations describing the growth of multicomponent systems and their approximate solutions must be compared with the Monte Carlo results in order to exclude nonphysical assertions and to prove the applicability of the used approximation scheme.

In Sect.2 the basic features of a unified microscopic interface theory of multicomponent crystal growth are discussed and their theoretical background (order parameter concept, master-equation approach, statistical approximation scheme, multilayered SOS-lattice interface model) is outlined briefly. In Sect. 3 basic master-equation approaches are represented in micro- and macro-variables. The equilibrium and non-equilibrium properties are discussed by means of several statistical approximation schemes. In Sect. 4 selected problems of two-component crystal growth (ordered growth and incorporation of impurities) are treated using the theoretical background described in Sects. 2 and 3. Finally, in Sect. 5 conclusions and future trends in the development of a microscopic theory of the growth of multicomponent crystals are given.

## 2 Basic Features

The general aim of a unified interface theory of crystal growth should be the description of the dynamics of the following elementary events on an atomistic length and microscopic time scale:

(a) fluid $\rightleftarrows$ solid (first-order) phase transitions, i.e. single-site one-particle attachment and detachment processes between fluid and solid states on the interface

and

(b) two-site two-particle exchange processes, i.e. mass transport (diffusion) processes within and between fluid and solid states of the interface.

Of course, an atomistic treatment of (a) and (b) on the frame of the full set of microscopic equations of motion is apparently impossible. On the other hand, we are usually only interested in the time evolution of the system on a coarse grained time scale relative to the inverse atomic frequencies of particles $m\alpha$ ($m = F, S$; $\alpha = A, B$) where the microscopic equations of motion may be replaced by a set of stochastic equations[26, 37, 40–42]. This means that one introduces so-called conditional (transition) probabilities for the realization of the elementary events (a) and (b) at the interface which can be characterized by local variables of interest. These local variables may be given as microvariables (e.g. site and height variables)[26–32, 44] or as macrovariables (e.g. particle numbers or concentrations)[43, 44], respectively. The probability of finding the system at any time in a micro- or macrostate, which is characterized by all the micro- or macrovariables of this state, should obey a so-called "master equation"[26, 37, 40–44]. All the interface dynamics enters into such a master equation through the transition probabilities of the elementary events (a) and (b). The latter can be constructed arbitrarily within the frame of the detailed balance condition[9, 21, 23, 40–45], accounting for different types of interface dynamics at the fluid-solid interface, e.g. to simulate the growth from the vapour, solution and melt, respectively[9, 10, 18, 21, 23, 39, 45]. The restriction imposed on a Markovian process means that the conditional (transition) probabilities of these elementary events are assumed to be independent of time and of the previous states. Using this assumption the interface kinetics (which describes the statistical cooperation of all these elementary events) is introduced as a stochastic process by a "Markovian master-equation approach". This approach is the starting point for a microscopic theory of multicomponent growth describing the elementary events (a) and (b) and their dynamics in terms of conditional (transition) probabilities. As order parameters (which specify the state of order of the system) we may take an appropriate set of space- and time-dependent macroscopic (averaged) concentration variables of particles which depict the SRO and/or LRO of particles within the multilayered SOS lattice-based fluid-solid interface. In[9, 10, 18, 20–23] the order parameters are determined according to different statistical approximation schemes:

(a) single-particle approximation (SPA)[9, 10, 18]

(b) two-particle (pair) approximation (PA)[9, 10, 18]

(c) Bragg-Williams approximation (BWA)[19–23].

The SPA includes only statistically independent events and provides no information about the spatial distribution of particles relative to each other but describes, of course, LRO. However, the PA describes the interaction within a pair exactly and the interactions with the remaining other neighbours only approximately so that the SRO and LRO are taken into account consistently[46, 47]. The BWA is characterized by the assumption of statistical independence (randomness) of the events. Therefore, any average (mean value) can be factorized self-consistently into products of single averages (see[48, 49]). In BWA only the LRO in the system can be included (SRO effects, e.g. the "interface roughness"[26, 28], have to be described by a long-range order parameter). Of course, it is expected that the used approximation scheme is appropriate if the statistical approximation is consistent with the range of the interaction between the particles. For instance, the BWA is correct for extremely long-range interactions whereas the PA treats the interaction within a pair exactly and is therefore more suitable for the description of systems with short-range interactions. Note that in the approximation schemes (a) and (c) a nucleation-controlled fluid $\leftrightarrows$ solid phase transition is excluded. In pair approximation, however, long-range correlations (like nucleation) can be comprehended but no further growth is possible if the nucleus contains more than two "lattice cells" (see[47]).

Finally, the applications and limitations of the multilayered SOS-lattice fluid-solid interface model are briefly described. In this SOS interface model the lattice sites are occupied either by a solid $S\alpha$ or fluid $F\alpha$ particle ($\alpha$ = A, B). These particles should be "classical particles" with internal degrees of freedom describing some "structural" distinctions between the fluid and solid state. The interaction energy between two particles should be composed of a hard core and an added soft (attractive) part. The hard-core constraint stipulates only a one-particle occupation per lattice site whereas the soft part is supposed to be an attractive short-range two-particle interaction (bond) energy between nearest neighbours in the lattice[29, 50]. Therefore, within a common lattice structure for the fluid and solid phase a $F\alpha$-particle differs from an $S\alpha$-particle due to different presupposed internal degrees of freedom and bond energies. However, from these findings no a priori assumptions about the real interface structure can be made. For this reason, several approximations have been made especially for a realistic modelling of the liquid (melt)-solid interface which has on its liquid and solid side about the same atomic density and SRO of particles (species) but a different LRO, due to a symmetry change at the interface. The former aspect can be roughly simulated within the frame of an SOS lattice-based description if the lattice sites at the interface are occupied by "solid-like ($S\alpha$)" and "liquid-like ($F\alpha$)" particles ($\alpha$ = A, B)[9, 19–23, 28, 73]. But the latter aspect, i.e. the symmetry change occurring in crystallization, and the atomic structure of the liquid near the interface can only be taken into account reliably by a molecular theory of a liquid-solid interface (see[51, 74–76]).

## 3 Basic Treatments

### 3.1 Master-Equation Approach in Microvariables

#### *3.1.1 Site Variables, Microstates and Hamiltonian of the System*

As already outlined in Sects. 1 and 2 we confine ourselves to an SOS lattice model of the two-component crystal growth, which represents a straightforward generalization of the one-component SOS lattice model (see e.g.[26–31, 38, 44, 52, 53]). According to this model each site $j = (x, y, z)$ of the s.c. lattice can be occupied by particles (species) $m\alpha$ of both phases $m\,(m = F, S)$ and components $\alpha\,(\alpha = A, B)$. Here, the integer $n \equiv z$ labels the layers parallel to the {100} interface, and the integer coordinates (x, y) define the position within such a layer. $n \to -\infty$ corresponds to the SOS-bulk solid and $n \to \infty$ to the SOS-bulk fluid phase. Furthermore, as site variables we can define occupation numbers $C_j^{m\alpha} = (1, 0)$ of the lattice site j with a particle $m\alpha$ belonging to the phase m and component $\alpha$. Thus, each microstate of the considered system is characterized by a set of site variables as microvariables $\{C_j^{m\alpha}\}$. Obviously, the following relations hold

$$\sum_{m=F,S} C_j^{m\alpha} = C_j^{\alpha}\,, \quad \sum_{\alpha=A,B} C_j^{m\alpha} = C_j^{m}\,,$$
$$\sum_{m=F,S} C_j^{m} = \sum_{\alpha=A,B} C_j^{\alpha} = 1\,, \qquad C_j^{m\alpha} = C_j^{m} C_j^{\alpha}\,, \tag{1}$$

where $C_j^{m}$ ($C_j^{\alpha}$) is the occupation number of the lattice site j with a particle belonging to the phase m (component $\alpha$). Taking into account only nearest neighbour interactions between particles ($m\alpha$), the Hamiltonian of the system can be written as

$$\hat{H} = \frac{1}{2} \sum_{\substack{j,\underline{k}\in j \\ m,\bar{m};\alpha,\bar{\alpha}}} \Phi^{m\alpha,\bar{m}\bar{\alpha}}\, C_j^{m\alpha} C_k^{\bar{m}\bar{\alpha}} + \sum_{j;m;\alpha} (\mathring{\mu}^{m\alpha} - 3\,\Phi^{m\alpha,m\alpha})\, C_j^{m\alpha} + \hat{V}\{C_j^{m}\} \tag{2}$$

$(m, \bar{m} = F, S;\ \alpha, \bar{\alpha} = A, B)$

where $\Phi^{m\alpha,\bar{m}\bar{\alpha}} < 0$ is the nearest neighbour interaction (bond) energy between the species ($m\alpha$) and ($\bar{m}\bar{\alpha}$); $k \in j$ means that k is a nearest neighbour of j, and j runs over all lattice sites. As well known, the SOS lattice model takes into account only configurational properties of the system (first term in Eq. (2)). The non-configurational part, which is connected with changes of the internal degrees of freedom of the species, lies outside the SOS lattice model and must be introduced ad hoc (second term in Eq. (2)) where $\mathring{\mu}^{m\alpha}(T, P)$ denotes the standard chemical potential of the species $m\alpha$. The third term in Eq. (2) takes into account the SOS conditions, $\hat{V} = \infty$, if the solid-fluid configuration $\{C_j^{m}\}$ contradicts the SOS restriction; otherwise, $\hat{V} = 0$[26, 52]. The Hamiltonian (Eq. (2)) contains only the relevant degrees of freedom which are connected with the order parameters describing the fluid $\leftrightarrows$ solid phase transition. In the following we also use a representation of microvariables, commonly applied in [7–18]:

$$C_j^{S} \equiv \eta_j\,, \qquad C_j^{\alpha} \equiv \delta_{\xi_j\alpha} \tag{3a}$$

where $\delta_{\xi_j\alpha}$ means the Kronecker symbol ($\delta_{\xi_j\alpha} = 1$ if $\xi_j = \alpha$ and equal zero if $\xi_j \neq \alpha$; $\alpha = A, B$). Now, a microstate of our system is given by the vector

$$g = \{\xi_1\eta_1\,;\ \xi_2,\eta_2\,;\ldots;\ \xi_j,\eta_j\,;\ldots\}$$
$$\xi_j = A, B\,; \qquad \eta_j = 0, 1\,. \tag{3b}$$

According to Eq. (3 a) and following[7–18], we also use the notation

$$m(= F, S) \rightarrow \eta (= 0, 1) ; \qquad \Phi^{m\alpha, \bar{m}\bar{\alpha}} \rightarrow \varphi_{\eta\bar{\eta}}^{\alpha\bar{\alpha}} ; \qquad \mathring{\mu}^{m\alpha} \rightarrow \mathring{\mu}_{\eta}^{\alpha} \tag{3c}$$

and introduce the entropy parameters $\Theta_{\eta}^{\alpha}$ by (see[7–9])

$$K_B T \Theta_{\eta}^{\alpha} = \mathring{\mu}_{\eta}^{\alpha}(T, P) - 3\varphi_{\eta\eta}^{\alpha\alpha} \tag{3d}$$

where $K_B$ is the Boltzmann constant, T the absolute temperature and P the pressure.

### 3.1.2 *Master-Equation for Elementary Events*

The considered system described by the Hamiltonian (Eq. (2)) cannot have any dynamical (relaxing) properties, because relation (2) commutes with the site variables $C_j^{m\alpha}$, i.e. their statistical averages do not depend on time (see[41, 54–56]). To obtain a dynamical model system, we have to add an external perturbation to Eq. (2) which describes the interaction with other degrees of freedom (e.g. heat bath)[37, 40, 41]. Then, starting from the Liouville equation of the whole system characterized by Eq. (2), the heat bath and the interaction Hamiltonian between the system described by Eq. (2) and the heat bath, a Markovian master-equation can be derived for the probability $\varrho(g, t)$ that a microstate g of the system (2) is realized at time t (see, e.g.[37, 40–42, 56, 57])

$$\frac{\partial}{\partial t}\varrho(g, t) = \sum_{g'} [W(g' \rightarrow g)\varrho(g', t) - W(g \rightarrow g')\varrho(g, t)] \tag{4a}$$

where, by definition,

$$\sum_{g} \varrho(g, t) = 1 \tag{4b}$$

holds. In principle, the W's can be determined deriving Eq. (4 a) from the Liouville equation of the whole system but mostly these are introduced as a stochastic process in terms of the conditional (transition) probabilities $W(g \leftrightarrows g')$ for the transitions $g \leftrightarrows g'$ between two microstates g and g'. The W's involve here all possible elementary events (attachment, detachment and diffusion processes of particles) and describe the interface dynamics. The interface kinetics is governed by the master-equation (4 a). Using Eq. (4 a) as starting point, the W's can be specified by the principle of detailed balance

$$\varrho_{eq}(g')W(g' \rightarrow g) = \varrho_{eq}(g)W(g \rightarrow g') \tag{4c}$$

so that the system is allowed to relax to the thermodynamic equilibrium state for long times $t \rightarrow \infty$, described by the canonical distribution function

$$\varrho_{eq} \sim \exp[-\hat{H}(g)/K_B T] \tag{4d}$$

where T is the absolute temperature of the heat bath. Combining Eqs. (4 c) and (4 d), we obtain

$$\frac{W(g \to g')}{W(g' \to g)} = \exp\left[(\hat{H}(g) - \hat{H}(g'))/K_B T\right] , \tag{5}$$

i.e. the transition probabilities are only defined as a ratio, and an arbitrary multiplier can be introduced. The absolute W's themselves lie outside our considerations and must be assumed ad hoc within the condition (5), i.e. there is still an infinite number of possibilities to select the W's which depict the interface dynamics. It is important that the W's in Eq. (5) derived for the equilibrium state can also be used for the nonequilibrium state. This is justified by the regression axiom of Onsager (see[39, 58]). Of course, the equilibrium properties are independent of the special choice of transition probabilities. Thus, Eqs. (2) and (4) define a stochastic model of crystal growth which can describe the equilibrium and nonequilibrium properties of the system.

Using representation (3) of microvariables, the transition probabilities for relaxation (one-particle single-site attachment and detachment) processes at the site i read

$$W(g \to g') = \delta_{\xi_1 \xi_1'} \delta_{\eta_1 \eta_1'} \dots \delta_{\xi_i \xi_i'} W^{\xi_i}_{\eta_i \eta_i'} \tag{6a}$$

where

$$W^{\xi_i}_{\eta_i \eta_i'} = W^{\xi_i}_{10} \delta_{\eta_i 1} \delta_{\eta_i' 0} + W^{\xi_i}_{01} \delta_{\eta_i 0} \delta_{\eta_i' 1} \tag{6b}$$

is the conditional probability of the transition of a particle $\xi_i = (A, B)$ from phase $\eta_i'$ to phase $\eta_i$. Analogously, for the two-particle two-site exchange process between two neighbouring sites j and i (diffusion process) we have

$$W(g \to g') = \delta_{\xi_1 \xi_1'} \delta_{\eta_1 \eta_1'} \dots \delta_{\xi_j \xi_j'} \delta_{\eta_j \eta_j'} W^{\xi_j \eta_j}_{\xi_i \eta_i} \dots \tag{7}$$

where $W^{\xi_j \eta_j}_{\xi_i \eta_i}$ is the conditional probability for the exchange of a particle $(\xi_i, \eta_i)$ at the site i with a particle $(\xi_j, \eta_j)$ at site j. According to Eqs. (2), (3) and (5), the transition probabilities fulfil the conditions[7–11, 13]

$$\frac{W^{\xi_i}_{01}}{W^{\xi_i}_{10}} = \exp\left[-\frac{1}{K_B T} \sum_{\nu, \eta'} 1^{\xi_i \nu}_{\eta'} (\varphi^{\xi_i \nu}_{0\eta'} - \varphi^{\xi_i \nu}_{1\eta'}) + \Theta_{\xi_i}\right] \tag{8}$$

with

$$\sum_{\nu, \eta'} 1^{\xi_i \nu}_{\eta'} = 6$$

for s.c. lattice where $\Theta_\xi = \Theta^\xi_1 - \Theta^\xi_0$ and $\eta' = 0, 1; \nu = A, B$. $1^{\xi_i \nu}_{\eta}$ is the number of nearest neighbours of the phase $\eta$ and the component $\nu$ to a particle of the component $\xi_i$ at site i. The $W^{\xi_j \eta_j}_{\xi_i \eta_i}$ fulfil similar conditions as Eq. (8) (see[9]). As already mentioned one of the transition probabilities in Eq. (5) must be chosen ad hoc. With respect to relaxation (8) we assume the attachment probability to be independent of the interface configuration, i.e. we put

$$W^{\nu}_{10} = \Omega_{\nu} \exp[-U_{\nu}|K_B T] \tag{9}$$

where $\Omega_\nu$ is an impingement frequency factor and $U_\nu$ an activation energy for the attachment of particle $\nu$ to the crystal (see[9, 10, 18]). Then, the detachment probability $W^{\xi_i}_{01}$ follows from Eq. (8). The SOS restriction is taken into account by appropriate prefactors which are zero if the process leads to a configuration being in contradiction to SOS:

$$W^{\xi_i}_{10} \to W^{\xi_i}_{10}\, \eta_{x,y,n-1}(1-\eta_{x,y,n}) \tag{10}$$

$$W^{\xi_i}_{01} \to W^{\xi_i}_{01}\, \eta_{x,y,n}(1-\eta_{x,y,n+1}) \tag{11}$$

with $j = (x, y, n)$.

Because of the $[0, \infty]$ property of the potential $\hat{V}$ in Eq. (2) and the corresponding [0, 1] property of the prefactors in Eqs. (10) and (11), the detailed balance conditions (5) are not violated. With respect to diffusion it is assumed for simplicity (see[9, 10, 18]) that exchange processes of particles $m\alpha \rightleftarrows \overline{m}\overline{\alpha}$ $(\alpha \neq \overline{\alpha} = A, B)$ between nearest neighbouring sites should be excluded if $m \neq \overline{m} = F, S$. This condition is realized by

$$W^{\xi_j \eta_j}_{\xi_i \eta_i} \to W^{\xi_j \eta}_{\xi_i \eta}\,(1-\delta_{\xi_j \xi_i})\, A_j A_i \prod_{k \in j,i} A_k \tag{12}$$

where $A_j = (1-\eta_j)$ stands for diffusion in the fluid phase $\eta = 0$ and $A_j = \eta_j$ for diffusion in the solid phase $\eta = 1$ within the interface, respectively (see[9]). Due to this simplification, the diffusion probabilities depend only on the mixing energy

$$\omega_\eta = \varphi^{AB}_{\eta\eta} - \frac{1}{2}\,(\varphi^{AA}_{\eta\eta} + \varphi^{BB}_{\eta\eta}) \tag{13}$$

in the phase $\eta = 0\,(F),\ 1\,(S)$. For ideal two-component systems is $\omega_\eta = 0$. Within Eqs. (5) and (12) the following choice of the absolute diffusion probability has been made[9, 10, 18]

$$W^{A\eta}_{B\eta} = \Omega_{\eta\eta} \exp\left[\frac{2(1^{AB}_\eta - 1)\,\omega_\eta}{K_B T}\right] \left\{\exp\left[\frac{2 1^{BB}_\eta\,\omega_\eta}{K_B T}\right] + \right. \tag{14}$$

$$\left. + \exp\left[\frac{2(1^{AB}_\eta - 1)\,\omega_\eta}{K_B T}\right]\right\}^{-1} = D_{\eta\eta}/a^2$$

where $\Omega_{\eta\eta}$ is a characteristic frequency factor for the particle exchange and $D_{\eta\eta}$ the corresponding diffusion coefficient in the phase $\eta = 0\,(F),\ 1\,(S)$ ("a" means the lattice constant). Note that according to Eq. (12) exchange processes between fluid and solid particles are excluded but, in principle, these can be taken into account (see[23, 44, 59]).

Now, using Eqs. (6) and (7), the master equation (4) turns to[7–9, 18)]

$$\frac{\partial}{\partial t}\varrho(\ldots;\xi_i,\eta_i;\xi_j,\eta_j;\ldots;t)=\sum_{i,\eta_i'} W^{\xi_i}_{\eta_i\eta_i'}\varrho(\ldots;\xi_i,\eta_i';\xi_j,\eta_j;\ldots;t)+$$

$$+\sum_{i,j\in i} W^{\xi_i\eta_j}_{\xi_j\eta_i}\varrho(\ldots;\xi_j,\eta_j;\xi_i,\eta_i;\ldots;t)-\left[\sum_{i,\eta_i'} W^{\xi_i}_{\eta_i'\eta_i}+\sum_{i,j\in i} W^{\xi_i\eta_i}_{\xi_j\eta_j}\right]\cdot \tag{15}$$

$$\cdot\,\varrho(\ldots;\xi_i,\eta_i;\xi_j,\eta_j;\ldots;t)\,.$$

The sum in Eq. (15) is performed over all neighbours j nearest to an i-site ($j\in i$), over all the i-sites of the lattice, and over their states $\eta_i' = 0\,(F), 1\,(S)$. Solutions of Eq. (15) can only be obtained in certain approximations. The most complete and detailed numerical analysis of two-component crystal growth is given by the Monte Carlo method where a direct stochastic simulation of the master equation (15) is performed in a finite system (see[11, 16–18, 38, 39)]). On the other hand, summation over a subset of microvariables in Eq. (15) leads to a hierarchy of interconnected evolution equations for the reduced probability distribution functions (PDF's) $\varrho^{(l)}$ ($l = 1, 2, \ldots$) with the following types of PDF's: (a) single-particle distribution function

$$\varrho^{(1)}(\xi_i,\eta_i;t)=\sum_{\substack{\ldots\xi_j,\eta_j\ldots\\(j\neq i)}}\varrho(g,t) \tag{16}$$

(b) two-particle (pair) distribution functions

$$\varrho^{(2)}(\xi_i,\eta_i;\xi_k,\eta_k;t)=\sum_{\substack{\ldots\xi_j,\eta_j\ldots\\(j\neq i,k)}}\varrho(g,t) \tag{17}$$

etc.

Thereby, the evolution equation for $\varrho^{(l)}$ ($l = 1, 2, \ldots$) contains PDF's of orders higher than l:

$$\frac{\partial}{\partial t}\varrho^{(l)}=f\{\ldots;\varrho^{(l)},\varrho^{(l+1)},\ldots;W^{\xi_i}_{\eta_i\eta_i'},W^{\xi_i\eta_i}_{\xi_j\eta_j}\}\,. \tag{18}$$

In order to obtain a closed set of equations, a statistical approximation scheme is necessary, which expresses the higher-order PDF's in terms of lower order ones (see e.g.[31, 46, 60)]). The simplest statistical approximation scheme resides in the single-particle approximation (SPA), i.e. we use the decoupling procedure:

$$\varrho^{(2)}(\xi_i,\eta_i;\xi_k,\eta_k;t)=\varrho^{(1)}(\xi_i,\eta_i;t)\,\varrho^{(1)}(\xi_k,\eta_k;t)\,. \tag{19}$$

In the two-particle (pair) approximation (PA) $\varrho^{(3)}, \varrho^{(4)}, \ldots$ are expressed by $\varrho^{(2)}, \varrho^{(1)}$, e.g.[60)]

$$\varrho^{(3)}(\xi_i,\eta_i;\xi_k,\eta_k;\xi_j,\eta_j;t)=\frac{\varrho^{(2)}(\xi_i,\eta_i;\xi_k,\eta_k;t)\,\varrho^{(2)}(\xi_k,\eta_k;\xi_j,\eta_j;t)}{\varrho^{(1)}(\xi_k,\eta_k;t)} \tag{20}$$

etc.

These statistical approximation schemes and also the BWA are used in the derivation of the kinetic equations for expectation values of microvariables.

### *3.1.3 Kinetic Equations*

In order to describe the kinetic properties of the considered system, kinetic (or rate) equations for macroscopic expectation values are to be derived within the SPA, PA and BWA. The macroscopic expectation value of any dynamic variable D (g) is defined in the usual way

$$\langle D(t)\rangle = \sum_{g} D(g)\,\varrho(g, t)\,. \tag{21}$$

Performing statistical averaging in Eq. (21) within the approximation scheme of the PDF $\varrho^{(1)}$ by means of Eq. (19), we obtain an expectation value in SPA. In addition, when using the approximation scheme of the PDF $\varrho^{(2)}$ together with Eq. (20), we get an average in PA. The BWA will be introduced later. Then, the time variation of $\langle D(t)\rangle$ reads together with Eq. (4)

$$\frac{d}{dt}\langle D(t)\rangle = -\left\langle \sum_{g'} [D(g) - D(g')]\,W(g \to g')\right\rangle\,. \tag{22}$$

Equation (22) is called kinetic or rate equation (see[9, 10, 18, 37, 57]) for the various moments of the quantity D (g) (e.g. of the occupation numbers in Eq. (1)).

According to Eq. (22) the average of the microvariable $C_j^{m\nu}$ of the species $m\nu$ ($m = F, S$; $\nu = A, B$) describing the concentration of a $\nu$-particle in the state m within any lattice site of the layer n is given by

$$\begin{aligned} \langle C_j^{m\nu}\rangle &\equiv C_\eta^\nu(n) \equiv C_n^{m\nu} = \sum_g C_j^{m\nu}\,\varrho(g, t) = \\ &= \sum_g A_j \delta_{\xi_j\nu}\,\varrho(g, t) = \sum_{\xi_j, \eta_j} A_j \delta_{\xi_j\nu}\,\varrho^{(1)}(\xi_j, \eta_j; t) \end{aligned} \tag{23}$$

where $A_j = \eta_j$ and $A_j = 1 - \eta_j$ stands for $\eta = 1\,(m = S)$ and $\eta = 0\,(m = F)$, respectively. "j" is here a lattice site belonging to the layer "n", so that $C_\eta^\nu(n)$ is an intralayer long-range order parameter which describes the LRO (i.e. the mean value of the concentration of the species $m\nu$ within the n-th layer) in the system. Analogously, we define concentrations characterizing the SRO within any layer and between neighbouring layers:

$$C_{\eta\eta'}^{\nu\nu'}(n) = \sum_{\xi_i, \eta_i; \xi_j, \eta_j} A_i A_j'\,\delta_{\xi_i\nu}\delta_{\xi_j\nu'}\,\varrho^{(2)}(\xi_i, \eta_i; \xi_j, \eta_j; t) \tag{24a}$$

and

$$C_{\eta\eta'}^{\nu\nu'}(n, n-1) = \sum_{\xi_i, \eta_i; \xi_j, \eta_j} A_i A_j'\,\delta_{\xi_i\nu}\delta_{\xi_j\nu'}\,\varrho^{(2)}(\xi_i, \eta_i; \xi_j, \eta_j; t) \tag{24b}$$

where $A_j' = \eta_j'$ for $\eta' = 1$ (m = S) and $A_j' = 1 - \eta_j'$ for $\eta' = 0$ (m = F). $C_{\eta\eta'}^{\nu\nu'}(n)$ is the concentration of pairs of neighbouring particles $(\nu, \eta)$ and $(\nu', \eta')$ within the layer n. $C_{\eta\eta'}^{\nu\nu'}(n, n-1)$ is the concentration of pairs of neighbouring particles $(\nu, \eta)$ and $(\nu', \eta')$, where the particle $(\nu, \eta)$ belongs to the layer n and $(\nu', \eta')$ to the layer n – 1 ($\nu, \nu' = A, B$; $\eta, \eta' = 0, 1$). The indices "j" in Eq. (23) and "i" in Eq. (24 a) denote sites in layer "n", the index "j" in Eq. (24 b) represents a site of layer "n – 1". Therefore, $C_{\eta\eta'}^{\nu\nu'}(n)$ has the meaning of an intralayer short-range order parameter and $C_{\eta\eta'}^{\nu\nu'}(n, n-1)$ of an interlayer short-range order parameter for the states $\eta$ and $\eta'$, respectively[9, 10, 18].

Carrying out the time derivative in Eq. (23), using the evolution equation (18) for l = 1 and taking advantage of the approximation (19), one arrives at the kinetic (rate) equation for $C_\eta^\nu(n)$ in single-particle approximation (SPA) of the form[9, 10, 18]:

$$\frac{d}{dt} C_\eta^\nu(n) = G^{(1)}\{C_\eta^\nu(n \pm \tilde{r})\}; \quad \tilde{r} = 1, 2 . \tag{25}$$

Using Eqs. (24), (18) and (20), we obtain in a similar way kinetic (rate) equations in pair approximation (PA) of the form[9, 10, 18]:

$$\frac{d}{dt} C_{\eta\eta'}^{\nu\nu'}\begin{pmatrix} n \\ n, n+1 \end{pmatrix} = \begin{matrix} G_I^{(2)} \\ G_{II}^{(2)} \end{matrix} \left\{ C_\eta^\nu(n \pm \tilde{r}),\, C_{\eta\eta'}^{\nu\nu'}(n \pm \tilde{r}),\, C_{\eta\eta'}^{\nu\nu'}(n \pm \tilde{r}, n \pm \tilde{r}') \right\} \tag{26}$$

$\tilde{r} = 1, 2$; $\tilde{r}' = 1, 2, 3$ .

The concentrations $C_\eta^\nu(n)$ and $C_{\eta\eta'}^{\nu\nu'}(n, n-1)$ are connected by[13, 14]

$$C_\eta^\nu(n) = \sum_{\eta', \nu'} C_{\eta\eta'}^{\nu\nu'}(n, n-1); \quad \nu, \nu' = A, B; \quad \eta, \eta' = 0, 1 . \tag{27}$$

Because of their space-consuming form we renounce an explicit representation of the functions $G^{(1)}$, $G_{I,II}^{(2)}$. They are given in[9, 10, 13, 14, 18]. In principle, kinetic equations in third (l = 3) – and higher – order approximations of the PDF's due to Eq. (18) can be derived. This is, however, more time-consuming. Furthermore, far from the interface, Eq. (25) or (26) with Eq. (27) reduces to the usual diffusion equation for $C_\eta^\nu(n)$ (written as finite-central difference equations)[9, 10, 13, 18]. Finally, other statistical approximations made for obtaining kinetic equations for the concentrations $C_\eta^\nu(n)$, $C_{\eta\eta'}^{\nu\nu'}(n), \ldots$ should be mentioned. As above, we form the time derivative of Eq. (24) using Eq. (18) for l = 2. However, before applying the PA (20), the transition probabilities in Eq. (18) are averaged according to Eq. (21), i.e. we put

$$W_{\eta\eta'}^{\xi} \to \langle W_{\eta\eta'}^{\xi} \rangle,\; W_{\beta\eta}^{\alpha\eta} \to \langle W_{\beta\eta}^{\alpha\eta} \rangle \tag{28}$$

where, in addition the averages of the exponentials in Eqs. (10), (11) and (12) are replaced by exponentials of the averages (see also[19, 18]). The resulting kinetic equations have the structure of Eq. (26), but now concentrations appear in the "exponentials" of the expressions of the transition probabilities due to Eq. (28). We arrive at a pair-like approximation with SRO similar as in the PA (26) but where the W's in Eq. (18) are replaced by their averages (28). We shall call this kind of statistical approximation of

Eq. (18) for l = 2 using Eqs. (20) and (28) as "mean-field pair approximation (MF-PA)" (for more details see[9, 10, 12, 13, 18]). Using the same procedure but within the single-particle approximation (19) one arrives at the Bragg-Williams approximation (BWA), which can also be obtained by a self-consistent truncation of Eq. (22) in lowest order (see Sect. 3.2). It can be shown that under the simplest conditions, i.e. neglecting diffusion in the solid ($\Omega_{11} = 0$) and assuming diffusion in the fluid to be infinitely fast ($\Omega_{00} \rightarrow \infty$), as well as for free miscibility in the fluid ($\omega_0 = 0$) and complete wetting ($\varphi_{10}^{\alpha\beta} = \varphi_{00}^{\alpha\beta}$), we obtain in BWA[9, 10]

$$\frac{d}{dt} C_1^\nu(n) = W_{10}^\nu \left\{ C_0^\nu[C_1(n-1) - C_1(n)] - \frac{C_1^\nu(n)}{C_1(n)} [C_1(n) - C_1(n+1)] \times \right. \tag{29}$$

$$\left. \times \exp\left[ \Theta_\nu - \frac{4\Phi_{\nu A} C_1^A(n) + 4\Phi_{\nu B} C_1^B(n)}{K_B T} - \frac{\Phi_{\nu A} C_1^A(n-1) + \Phi_{\nu B} C_1^B(n-1)}{C_1(n-1) K_B T} \right] \right\}$$

where $C_0^\nu = C_0^\nu(n)/(1 - C_1(n)) = \text{const.}$ is the concentration of the component $\nu$ in the fluid phase ($\eta = 0$), $C_1(n)$ the concentration of the solid ($\eta = 1$) in the layer n, $\Phi_{\nu\gamma} = \varphi_{01}^{\nu\gamma} - \varphi_{11}^{\nu\gamma}$ and $\Theta_\nu = (3\Phi_{\nu\nu} - \Delta\mathring{\mu}_\nu)/K_B T \, (\Delta\mathring{\mu}_\nu = \mathring{\mu}_0^\nu - \mathring{\mu}_1^\nu)$. If the layer concentration $C_1^\nu(n)$ is determined within one of the above-mentioned statistical approximation schemes (SPA, PA, MF-PA, BWA), a dimensionless interfacial velocity (growth rate) can be calculated by

$$R = \sum_{\nu, n} (W_{10}^\nu)^{-1} \frac{d}{dt} C_1^\nu(n) \, . \tag{30}$$

The state of equilibrium is here defined by R = 0. Thus, the kinetic Eqs. (25), (26) or (29) together with (30) form a closed set of equations which describes the growth (interface) kinetics of the two-component system in SPA, PA or BWA, respectively. The MF-PA can be obtained with the above-mentioned procedure.

In Sect. 3.1.4 general results for binary crystal growth are discussed within the frame of the PA and MF-PA. A comparison with the Monte Carlo simulation results is given.

### *3.1.4 General Results on Binary Crystal Growth*

The kinetic equations discussed in Sect. 3.1.3 represent nonlinear difference-differential equations which are solved numerically for a finite number of layers. As result one gets equilibrium (R = 0) and kinetic ($R \neq 0$) phase diagrams (solidus $C_1^A(n = -\infty, T)$ and liquidus $C_0^A(n = \infty, T)$), the growth rate R as a function of the deviation from equilibrium and the long- and short-range order parameters (23) and (24) of the two-component system[7–18]. These properties depend on the following entry parameters: kinetic parameters $W_{10}^\alpha$, $\Omega_{\eta\eta}$, energy parameters $\varphi_{\eta\eta'}^{\alpha\beta}$ and thermodynamic parameters $\Theta_\alpha$. A comparison with the Monte Carlo simulation shows that the "best" statistical approximation scheme is the pair approximation (PA) whereas the MF-PA gives unsatisfactory agreement with the results of the computer simulation[9, 10, 12, 18] (see Fig. 1). Because of the good agreement between the Monte Carlo results and the results in PA, higher-order approximations are not necessary. Of course, the SPA and BWA fail, if short-range order is essential.

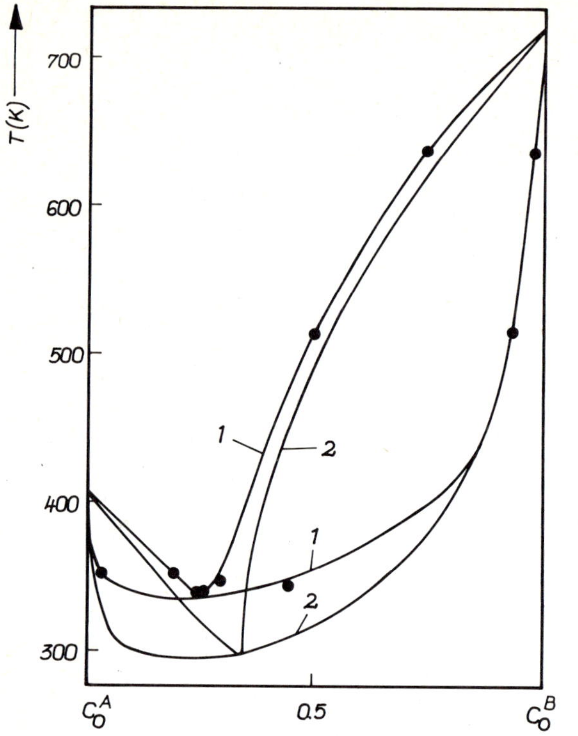


**Fig. 1.** Kinetic phase diagram for $R = 0.1$ with the entry parameters $W_{B0}^{A0} | W_{10} \rightarrow \infty$, $\Phi_{AA} = 1254$, $\Phi_{BB} = 6270$ and $\Phi_{AB} = 830$ J/mol; $\Theta_A = 1$, $\Theta_B = 3\,(W_{10}^A = W_{10}^B \equiv W_{10})$: (1) corresponds to the pair approximation (PA) and (2) to the mean-field pair approximation (MF-PA). Solid circles ● mark the Monte Carlo simulation results (after Cherepanova et al.[9, 10])

All the approximations used (SPA, PA, MF-PA, BWA) introduce artificial metastable states with infinite lifetimes (i.e. $R = 0$ for supersaturations smaller than a critical one). However, for atomically rough interfaces considered throughout, this critical region is negligible (see[9, 10, 13–15]). In order to present concrete results one has to reduce the space of entry parameters. A summary of the general growth features of binary crystals is given in[10–18, 45]. Here, we will discuss only the results reported in[10] where it is assumed that $W_{10}^A = W_{10}^B \equiv W_{10}$, $\Omega_{11} = 0$ (no diffusion in the solid), $\Omega_{00} = \infty$ (infinitely fast diffusion in the fluid) and that complete wetting occurs ($\varphi_{01}^{\alpha\beta} = \varphi_{00}^{\alpha\beta}$). As is well known, the energy parameters $\Phi_{\alpha\beta} = \varphi_{01}^{\alpha\beta} - \varphi_{11}^{\alpha\beta}$ determine the topology of the phase diagram. For $\Phi_{AA} < \Phi_{AB} < \Phi_{BB}$, the phase diagrams exhibit a cigar-like shape. For $\Phi_{AB} > \Phi_{AA}, \Phi_{BB}$, we obtain phase diagrams with a maximum. In this case, the solid tends to display an A-B superstructure. For $\Phi_{AB} < \Phi_{AA}, \Phi_{BB}$ phase diagrams with a minimum result, leading to eutectic phase diagrams if $\Phi_{AA} - \Phi_{AB}$ and $\Phi_{BB} - \Phi_{AB}$ increase. In this case, the components in the solid tend to undergo clustering. The variation of the growth rate with temperature for different compositions of the bulk fluid phase is shown in Fig. 2 for $\Phi_{AB} > \Phi_{AA}, \Phi_{BB}$. As in the case of eutectic alloys ($\Phi_{AB} < \Phi_{AA}, \Phi_{BB}$), the function $R(T)$ is linear for pure components (normal growth mode) and nonlinear in between (note that in Fig. 2 only small kinetic undercoolings $\Delta T = T_{eq} - T$ are resonable, $T_{eq}$ being the equilibrium temperature of the system determined by $R(T = T_{eq}) = 0$). Evaluating Eq. (26) for $\Omega_{11} = 0$ (no diffusion in the solid), the influence of the diffusion process in the fluid ($\Omega_{00}$ = finite) on two-component crystal growth has been examined and discussed in[12, 13, 18]. Furthermore, with respect to realistic

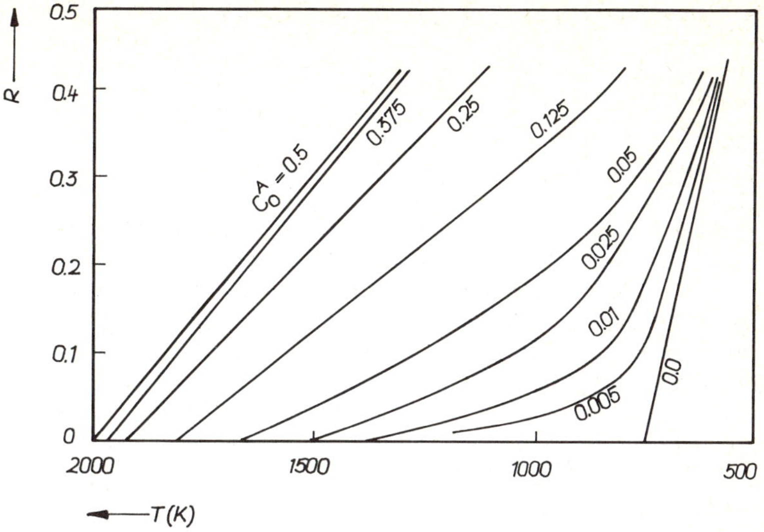


**Fig. 2.** Dependence of the growth rate R on temperature T for different compositions of the liquid phase $C_0^A$ with the entry parameters $W_{B0}^{A0} | W_{10} \to \infty$; $\Phi_{AA} = \Phi_{BB} = 6285$, $\Phi_{AB} = 22\,836$ J/mol; $\Theta_A = \Theta_B = 3$ (after Cherepanova et al[9, 10]) in pair approximation

situations, the kinetic equation approach in pair approximation (26) has been extended to more complicated lattices (b.c.c., f.c.c., diamond) with and without overhangs at the solid-fluid interface[13–15, 18]. It seems that the model without overhangs fits better a real physical system (e.g. Si-Ge alloy). A survey of the discussed topics (binary alloys[15, 16], eutectic alloys[11, 12, 17] and binary crystals with a cigar-shaped kinetic phase diagram (containing a maximum)[10]) is given in Table 1 (see also Sect. 5), which also includes the growth of ordered binary crystals with ionic-type interactions studied in[3, 4, 24, 25, 45, 61, 62, 68] (see also Sect. 4.1).

**Table 1.** Subjects discussed in this article

| Crystal structure | Main subject | Ref. |
|---|---|---|
| binary chain | kink growth of mixed crystals (from vapour, solution) | 2, 3, 5, 6 |
| binary Kossel (100) | melt growth of regular alloys, eutectic alloys, binary crystals with cigar-shaped (with a maximum or minimum) kinetic phase diagram | 10–12, 15–18, 23 |
| binary Kossel (100) | vapour, solution and melt growth of ordered AB crystals; kinetic disordering | 3, 4, 24, 25, 45, 61, 62, 68 |
| binary crystals: bcc: (100), (110) fcc: (100), (111) diamond structure: (100), (110), (111) | growth rate of (hkl) faces with and without overhangs (melt growth) | 14, 15, 18 |
| binary Kossel (100) | impurity incorporation; segregation (vapour and solution growth) | 3, 27, 69–71 |

## 3.2 Master-Equation Approach in Macrovariables

### *3.2.1 Macrovariables, Macrostates and Free Energy of the System*

In this kind of master-equation approach, we will use instead of microvariables "macrovariables" of the system right from the beginning. A macrovariable[43] is defined as an extensive ($\tilde{Y}$) or normalized intensive ($\tilde{y} = \tilde{Y}/\Omega$) quantity describing the "macrostates" $\{\tilde{Y}\}$ or $\{\tilde{y}\}$ of a large system (the size of which is $\Omega$) composed of a great number of elementary units (particles) interacting with each other and possibly with the environment (heat bath). In our considerations $\tilde{Y}$ corresponds to the number of particles and $\tilde{y}$ to its concentration. In constrast to a "microvariable", a "macrovariable" is a sum-function composed of a great number of microvariables the statistical expectation value of which is usually considered as a value realized in a given "macrostate". We characterize a macrostate $\{\tilde{Y}\}$ or $\{\tilde{y}\}$ of a system by its macrovariables ($\tilde{Y}$) or ($\tilde{y}$) and its Helmholtz free energy $\tilde{F}[(\tilde{Y})$ or $(\tilde{y})]$. To include fluctuations we regard a "macrovariable" as a "stochastic (random) variable". Because any macrostate describes also a state of order[48] the cluster variation method (CVM)[49] appropriately includes SRO and LRO effects in the system free energy. The CVM is a hierarchy of closed form approximations[63] of cooperative phenomena which comprises the Bragg-Williams approximation (BWA)[20–23], the quasi-chemical approximation (QCA)[25, 44] and higher-order approximations[48, 49]. The CVM can also be extended to kinetic problems in the frame of the path probability method (PPM)[48, 49]. Here, we restrict ourselves to the simplest case of the CVM, the BWA, and consider the multilayer SOS-lattice model of the fluid-solid interface, described in Sect. 3.1, within this statistical approximation scheme. Neglecting preferential clustering and SRO effects, i.e. assuming a random distribution of species m$\alpha$ in each layer "n" we obtain a BWA characterizing the random intralayer distribution of species (m = F, S; $\alpha$ = A, B). In this intralayer BWA, the Cowley-Warren SRO parameter[48] $\alpha_n (= 1 - \tilde{N}_n^{m\alpha}/N\tilde{C}_n^{m\alpha})$ is equal to zero, i.e. it yields

$$\tilde{C}_n^{m\alpha} = \tilde{N}^{m\alpha}/N = \frac{1}{N}\sum_{jvn} C_j^{m\alpha}\ ; \qquad n = 0, \pm 1, \pm 2, \ldots \tag{31}$$

where $\tilde{N}_n^{m\alpha}$ is the number and $\tilde{C}_n^{m\alpha}$ the concentration of species m$\alpha$ in the n-th layer. $C_j^{m\alpha}$ is defined by Eq. (1) where "jvn" means that the sum (31) runs only over the x- and y-coordinates of particles within the layer "n". $\tilde{N}_n^{m\alpha}$ is an extensive and $\tilde{C}_n^{m\alpha}$ the corresponding normalized intensive macrovariable which are considered as stochastic (random) variables. N denotes the total (fixed) number of species m$\alpha$ in each layer. Thus, any macrostate $\{\tilde{C}_n^{m\alpha}\}$ of the system is characterized by the set of the macrovariables

$$(\tilde{C}_n^{m\alpha}) = (\tilde{C}_n^{FA}, \tilde{C}_n^{SA}, \tilde{C}_n^{FB}, \tilde{C}_n^{SB}) \tag{32}$$

with $\alpha$ = A, B; n = 0, ±1, ±2, .... Moreover, the $\tilde{C}_n^{m\alpha}$ fulfil, due to Eq. (31), also the law of conservation of the number of species (1), i.e.

$$\sum_{m=F,S} \tilde{C}_n^{m\alpha} = \tilde{C}_n^{\alpha}, \quad \sum_{\alpha=A,B} \tilde{C}_n^{m\alpha} = \tilde{C}_n^{m}, \quad \sum_{m=F,S}\ \sum_{\alpha=A,B} \tilde{C}_n^{m\alpha} = 1 \tag{33}$$

where $\tilde{C}_n^m$ ($\tilde{C}_n^\alpha$) is the concentration of particles in the layer "n" belonging to the state m (component $\alpha$). The representation of Eq. (32) is advantageous if the intralayer diffusion is assumed to be infinitely fast[22, 23, 44]. Due to Eq. (33) three of the four concentrations in Eq. (32) are independent. We can e.g. write[19, 22, 23]:

$$\tilde{C}_n^{SA} = \tilde{x}_n \tilde{C}_n,\ \tilde{C}_n^{FA} = \tilde{\eta}_n (1 - \tilde{C}_n),\ \tilde{C}_n^{SB} = (1 - \tilde{x}_n) \tilde{C}_n,\ \tilde{C}_n^{FB} = (1 - \tilde{\eta}_n)(1 - \tilde{C}_n) \tag{34}$$

where the following notations are used: $\tilde{\eta}_n^A \equiv \tilde{\eta}_n$, $\tilde{\eta}_n^B = 1 - \tilde{\eta}_n$, $\tilde{x}_n^A \equiv \tilde{x}_n$, $\tilde{x}_n^B = 1 - \tilde{x}_n$, $\tilde{C}_n^S \equiv \tilde{C}_n$ and $\tilde{C}_n^F = 1 - \tilde{C}_n$. Here, $\tilde{x}_n^\alpha$ ($\tilde{\eta}_n^\alpha$) denotes the fraction of the component "$\alpha$" in the solid (fluid) state within the layer "n". The description of the interface can be improved if we distinguish between "free (f)" and "covered (c)" solid and fluid particles in each layer[20]. Covered solid (fluid) particles have two solid (fluid) neighbours and free solid (fluid) particles have a fluid (solid) neighbour normal to the growing plane. Accordingly, we have to introduce the concentration of free and covered solid or fluid $\alpha$-particles $\overset{\tilde{f}}{x}{}_n^\alpha$, $\overset{\tilde{c}}{x}{}_n^\alpha$, $\overset{\tilde{f}}{\eta}{}_n^\alpha$, $\overset{\tilde{c}}{\eta}{}_n^\alpha$, respectively. According to[20] it holds:

$$\tilde{C}_n^{S\alpha} = \overset{\tilde{c}}{x}{}_n^\alpha \tilde{C}_{n+1} + \overset{\tilde{f}}{x}{}_n^\alpha (\tilde{C}_n - \tilde{C}_{n+1}) = \tilde{x}_n^\alpha \tilde{C}_n$$

$$\tilde{C}_n^{F\alpha} = \overset{\tilde{c}}{\eta}{}_n^\alpha (1 - \tilde{C}_{n-1}) + \overset{\tilde{f}}{\eta}{}_n^\alpha (\tilde{C}_{n-1} - \tilde{C}_n) = (1 - \tilde{C}_n) \tilde{\eta}_n^\alpha \tag{35}$$

where $\overset{\tilde{p}}{x}{}_n^A \equiv \overset{\tilde{p}}{x}{}_n = 1 - \overset{\tilde{p}}{x}{}_n^B$, $\overset{\tilde{p}}{\eta}{}_n^A \equiv \overset{\tilde{p}}{\eta}{}_n = 1 - \overset{\tilde{p}}{\eta}{}_n^B$ ($p = c, f$). Now, we consider a system where "N" and the pressure "P" are constant and the dependence of the standard chemical potential $\mathring{\mu}^{m\alpha}$ of each species $m\alpha$ on T and P is known. Then, the free energy $\tilde{F}$ of a multilayered SOS lattice-based fluid-solid interface of a two-component system has formally in BWA the following functional structure:

$$\tilde{F} = \begin{cases} \tilde{F}[\tilde{C}_n, \tilde{x}_n, \tilde{\eta}_n; \Phi^{m\alpha, \bar{m}\bar{\alpha}}; T; \mathring{\mu}^{m\alpha}(T)] \quad {}^{22, 23)} & (36\,a) \\ \tilde{F}[\tilde{C}_n, \overset{\tilde{f}}{x}_n, \overset{\tilde{c}}{x}_n, \overset{\tilde{f}}{\eta}_n, \overset{\tilde{c}}{\eta}_n; \Phi^{m\alpha, \bar{m}\bar{\alpha}}; T; \mathring{\mu}^{m\alpha}(T)] \quad {}^{20)}\,. & (36\,b) \end{cases}$$

In Eq. (36) the "solid state (S)" and the "fluid state (F)" are introduced ad hoc (as discussed in Sects. 1 and 2) whereas in other approaches (see e.g.[64]) the existence of a coarse-grained free-energy functional is assumed to describe two coexisting phases for any component (see[41, 65]). In the following, we will use the representation (36 a) of the system free energy and discuss Eq. (36 b) in Sect. 4.2.

### *3.2.2 Master-Equation and Elementary Events*

The Markovian master-equation for the probability $P[\tilde{y}; t] = 1/N\ P[\tilde{Y}; t]$ of finding the system in contact with a heat bath of temperature T in a macrostate $\{\tilde{y}\}$ specified by the macrovariables $(\tilde{y}) = (\tilde{Y}/N)$ at time t is[41, 43, 44]

$$\frac{1}{N} \frac{\partial}{\partial t} P[\tilde{y}; t] = -\sum_r \left[ \tilde{W}\left(\tilde{y} \to \tilde{y} + \frac{r}{N}\right) P[\tilde{y}; t] - \tilde{W}\left(\tilde{y} + \frac{r}{N} \to \tilde{y}\right) P\left[\tilde{y} + \frac{r}{N}; t\right]\right] \tag{37}$$

where $\tilde{W}(\tilde{y} \rightarrow \tilde{y} + r/N)$ is the conditional probability per unit time for the transition from $\tilde{y}$ to $\tilde{y} + r/N$ with elementary jumps $r = \pm 1$. Into the $\tilde{W}$'s the following elementary events are included:

$$\tilde{W}\left(\tilde{y} \rightarrow \tilde{y} + \frac{r}{N}\right) = \begin{cases} \tilde{W}_R\left[\tilde{C}_n^{\bar{m}\bar{\alpha}}, \tilde{C}_n^{\bar{\bar{m}}\bar{\alpha}} \rightarrow \tilde{C}_n^{\bar{m}\bar{\alpha}} + \frac{r}{N}, \tilde{C}_n^{\bar{\bar{m}}\bar{\alpha}} - \frac{r}{N}\right] & \text{(38 a)} \\ \tilde{W}_D\left[\tilde{C}_n^{\bar{m}\bar{\alpha}}, \tilde{C}_{n'}^{\bar{m}\bar{\alpha}}; \tilde{C}_{n'}^{\bar{\bar{m}}\bar{\bar{\alpha}}}, \tilde{C}_n^{\bar{\bar{m}}\bar{\bar{\alpha}}} \rightarrow \tilde{C}_n^{\bar{m}\bar{\alpha}} + \frac{r}{N}, \right. & \\ \left. \tilde{C}_{n'}^{\bar{m}\bar{\alpha}} - \frac{r}{N}, \tilde{C}_{n'}^{\bar{\bar{m}}\bar{\bar{\alpha}}} - \frac{r}{N}, \tilde{C}_n^{\bar{\bar{m}}\bar{\bar{\alpha}}} + \frac{r}{N}\right]. & \text{(38 b)} \end{cases}$$

Equation (38 a) represents the conditional probability per unit time for the transition from the macrostate $\{\tilde{C}_n^{\bar{m}\bar{\alpha}}, \tilde{C}_n^{\bar{\bar{m}}\bar{\alpha}}\}$ to the macrostate $\{\tilde{C}_n^{\bar{m}\bar{\alpha}} + r/N, \tilde{C}_n^{\bar{\bar{m}}\bar{\alpha}} - r/N\}$ for $\bar{\alpha} = A, B$; $\bar{m} \neq \bar{\bar{m}} = F, S$; $r = \pm 1$; $n = 0, \pm 1, \pm 2, \ldots$ and decribes one-particle attachment $(F\bar{\alpha}) \rightarrow (S\bar{\alpha})$ and detachment $(S\bar{\alpha}) \rightarrow (F\bar{\alpha})$ processes in the layer "n". Eq. (38 b) denotes the conditional probability per unit time for the transition from the macrostate $\{\tilde{C}_n^{\bar{m}\bar{\alpha}}, \tilde{C}_{n'}^{\bar{m}\bar{\alpha}}; \tilde{C}_{n'}^{\bar{\bar{m}}\bar{\bar{\alpha}}}, \tilde{C}_n^{\bar{\bar{m}}\bar{\bar{\alpha}}}\}$ to the macrostate $\{\tilde{C}_n^{\bar{m}\bar{\alpha}} + r/N, \tilde{C}_{n'}^{\bar{m}\bar{\alpha}} - r/N; \tilde{C}_{n'}^{\bar{\bar{m}}\bar{\bar{\alpha}}} - r/N, \tilde{C}_n^{\bar{\bar{m}}\bar{\bar{\alpha}}} + r/N\}$ for $\bar{\alpha} \neq \bar{\bar{\alpha}} = A, B$; $\bar{m} = \bar{\bar{m}} = F, S$ or $\bar{\alpha}, \bar{\bar{\alpha}} = A, B$; $\bar{m} \neq \bar{\bar{m}} = F, S$ with $n \neq n' = 0, \pm 1, \pm 2, \ldots$; $r = \pm 1$ and simulates the two-particle exchange (diffusion) process of species $\bar{m}\bar{\alpha}$ and $\bar{\bar{m}}\bar{\bar{\alpha}}$ from layer "n" to a neighbouring layer "n′" and vice versa. In principle an intralayer diffusion can also be taken into account (which affects the distribution in a single layer) but it is neglected in the used intralayer BWA.

The $\tilde{W}$'s in Eq. (38) can be chosen arbitrarily apart from the fact that they must satisfy a detailed-balance condition[41, 43]:

$$P_{eq}[\tilde{y}]\,\tilde{W}\left(\tilde{y} \rightarrow \tilde{y} + \frac{r}{N}\right) = P_{eq}\left[\tilde{y} + \frac{r}{N}\right]\tilde{W}\left(\tilde{y} + \frac{r}{N} \rightarrow \tilde{y}\right) \tag{39}$$

where $P_{eq}[\tilde{y}]$ is the statistical distribution of a normalized intensive macrovariable $\tilde{y}$ at equilibrium. It holds[40, 43]

$$P_{eq}[\tilde{y}] = \exp\left[-\frac{1}{N}\tilde{F}[\tilde{y}]/K_BT\right] \Big/ \sum_{\{\tilde{y}\}} \exp\left[-\frac{1}{N}\tilde{F}[\tilde{y}]/K_BT\right] \tag{40}$$

where $\tilde{F}[\tilde{y}]$ means the system free energy in the macrostate $\{\tilde{y}\}$. Note that in Eq. (40) $\tilde{F}[\tilde{y}]$ appears instead of $\hat{H}(g)$ in Eq. (4 d). Calling to mind that a macrostate is composed of a great number of microstates which can be expressed in terms of the entropy of the system $S[\tilde{y}]$ by $\exp[S[\tilde{y}]/K_BN]$ where $\tilde{F}[\tilde{y}] = \tilde{E}[\tilde{y}] - TS[\tilde{y}]$ ($\tilde{E}[\tilde{y}]$ is the internal energy of the system in the macrostate $\{\tilde{y}\}$ and corresponds to the average of $\hat{H}$ over all microstates g connected with this macrostate). The expectation value of a macrovariable $D[\tilde{y}]$ at time t is defined as[43, 66)]

$$\langle D[\tilde{y}; t]\rangle = \sum_{\{\tilde{y}\}} D[\tilde{y}]\,P[\tilde{y}; t] \tag{41 a}$$

where the sum in Eq. (41 a) runs over all distinguishable accessible macrostates $\{\tilde{y}\}$ of the system. In general, we will use the following notation:

$$\langle \tilde{O} \rangle = O \tag{41b}$$

where $O = C_n^{m\alpha}, x_n^\alpha, \overset{p}{x}{}_n^\alpha, \eta_n^\alpha, \overset{p}{\eta}{}_n^\alpha$ (p = c, f) etc. are averaged (macroscopic) concentration variables in the representations (34) and (35), respectively.

In the following subsection the rate equations for expectation values of macrovariables are derived and discussed.

### 3.2.3 Rate Equations

Using the representation (34) of the macrovariables it follows from Eqs. (41), (37) and (38) that the time evolution of the expectation value $\langle D[\tilde{C}_n^{m\alpha}; t]\rangle$ is governed by the rate equation[37, 57]

$$\frac{1}{N}\frac{\partial}{\partial t}\langle D[\tilde{C}_n^{m\alpha}; t]\rangle = \sum_{\{\tilde{C}_n^{m\alpha}\}} D[\tilde{C}_n^{m\alpha}]\frac{\partial}{\partial t} P[\tilde{C}_n^{m\alpha}; t] =$$

$$= -\Bigg\langle \sum_{r=\pm 1}\Bigg\{ D[\tilde{C}_n^{m\alpha}] - D\Big[\tilde{C}_n^{m\alpha} + \frac{r}{N}\Big]\Bigg\}\Bigg\{ \sum_{\bar{m}(\neq m)=F,S} \tilde{W}_R\Big(\tilde{C}_n^{m\alpha}, \tilde{C}_n^{\bar{m}\alpha} \to$$

$$\tilde{C}_n^{m\alpha} + \frac{r}{N}, \tilde{C}_n^{\bar{m}\alpha} - \frac{r}{N}\Big) + \sum_{n'}^{(n)}\Big[ \sum_{\bar{m}(\neq m)=F,S}\ \sum_{\bar{\alpha}=A,B} \tilde{W}_D\Big(\tilde{C}_n^{m\alpha}, \tilde{C}_{n'}^{m\alpha}; \tilde{C}_{n'}^{\bar{m}\bar{\alpha}}, \tilde{C}_n^{\bar{m}\bar{\alpha}} \to \tag{42}$$

$$\tilde{C}_n^{m\alpha} + \frac{r}{N}, \tilde{C}_{n'}^{m\alpha} - \frac{r}{N}; \tilde{C}_{n'}^{\bar{m}\bar{\alpha}} - \frac{r}{N}, \tilde{C}_n^{\bar{m}\bar{\alpha}} + \frac{r}{N}\Big) + \sum_{\bar{\alpha}(\neq\alpha)=A,B} \tilde{W}_D\Big(\tilde{C}_n^{m\alpha}, \tilde{C}_{n'}^{m\alpha}; \tilde{C}_{n'}^{m\bar{\alpha}},$$

$$\tilde{C}_n^{m\bar{\alpha}} \to \tilde{C}_n^{m\alpha} + \frac{r}{N}, \tilde{C}_{n'}^{m\alpha} - \frac{r}{N}; \tilde{C}_{n'}^{m\bar{\alpha}} - \frac{r}{N}, \tilde{C}_n^{m\bar{\alpha}} + \frac{r}{N}\Big)\Big]\Bigg\}\Bigg\rangle$$

($\alpha$ = A, B; m = F, S; n = 0, ±1, ±2, ...). The summation $\sum_{n'}^{(n)}$ runs over nearest neighbouring layers (n' = n + l', l' = ±1) with respect to layer "n". Equation (42) is not a closed equation for $\langle D[\tilde{C}_n^{m\alpha}; t]\rangle$ but involves higher moments of $D[\tilde{C}_n^{m\alpha}]$ as well, i.e. we obtain a hierarchy of interconnected equations including correlations, $\langle \tilde{C}_n^{m\alpha}\tilde{C}_{\bar{n}}^{\bar{m}\bar{\alpha}}\rangle$, and fluctuations, $\langle(\tilde{C}_n^{m\alpha} - C_n^{m\alpha})^2\rangle$, in the macrovariables $\tilde{C}_n^{m\alpha}$ for all times[43, 57, 66]. According to Eq. (41) we have

$$C_n^{m\alpha} = \langle \tilde{C}_n^{m\alpha}(t)\rangle = \sum_{\{\tilde{C}_n^{m\alpha}\}} \tilde{C}_n^{m\alpha} P[\tilde{C}_n^{m\alpha}; t] \tag{43}$$

where $C_n^{m\alpha}$ is the macroscopic concentration of species m$\alpha$ in the layer "n" at time t and has the meaning of an intralayer LRO parameter (in the nonequilibrium state). Neglecting correlation and fluctuation effects in the order parameters, i.e. $\langle \tilde{C}_n^{m\alpha}\tilde{C}_{\bar{n}}^{\bar{m}\bar{\alpha}}\rangle = C_n^{m\alpha}C_{\bar{n}}^{\bar{m}\bar{\alpha}}$

and $\langle(\tilde{C}_n^{m\alpha})^2\rangle = (C_n^{m\alpha})^2$, Eq. (42) can be truncated self-consistently using the approximation $\langle f[\tilde{O}]\rangle \to f[\langle\tilde{O}\rangle]$ in BWA with the average $\langle\tilde{O}\rangle$ defined by Eq. (43). Under these additional assumptions, we obtain from Eq. (42) the following rate equation for the time evolution of $\langle D[\tilde{C}_n^{m\alpha}; t]\rangle = C_n^{m\alpha}$ in BWA[19, 22, 23]:

$$\frac{d}{dt} C_n^{m\alpha} = \sum_{\bar{m}(\neq m)=F,S} [K_n^{\bar{m}\alpha \to m\alpha} - K_n^{m\alpha \to \bar{m}\alpha}] + \sum_{\bar{\alpha}(\neq \alpha)=A,B} \sum_{l'=\pm 1} [K_{n+l' \to n}^{m\bar{\alpha} \to m\alpha} - K_{n \to n+l'}^{m\alpha \to m\bar{\alpha}}] +$$

$$+ \sum_{\bar{\alpha}=A,B} \sum_{\bar{m}(\neq m)=F,S} \sum_{l'=\pm 1} [K_{n+l' \to n}^{\bar{m}\bar{\alpha} \to m\alpha} - K_{n \to n+l'}^{m\alpha \to \bar{m}\bar{\alpha}}] ; \qquad (44)$$

$$\alpha = A, B; \quad m = F, S; \quad n = 0, \pm 1, \pm 2, \ldots$$

where

$$K_n^{\bar{m}\alpha \rightleftarrows m\alpha} = W_R\left[C_n^{m\alpha}, C_n^{\bar{m}\alpha} \to C_n^{m\alpha} + \frac{r}{N}, C_n^{\bar{m}\alpha} - \frac{r}{N}\right] \qquad (45\,a)$$

$(\alpha = A, B; \quad m \neq \bar{m} = F, S; \quad n = 0, \pm 1, \pm 2, \ldots)$

and

$$K_{n+l' \rightleftarrows n}^{\bar{m}\bar{\alpha} \rightleftarrows m\alpha} = W_D\left[C_n^{m\alpha}, C_{n+l'}^{m\alpha}; C_{n+l'}^{\bar{m}\bar{\alpha}}, C_n^{\bar{m}\bar{\alpha}} \to C_n^{m\alpha} + \frac{r}{N}, C_{n+l'}^{m\alpha} - \frac{r}{N}; C_{n+l'}^{\bar{m}\bar{\alpha}} - \frac{r}{N}, C_n^{\bar{m}\bar{\alpha}} + \frac{r}{N}\right] \qquad (45\,b)$$

($\bar{\alpha} \neq \alpha = A, B; m = \bar{m} = F, S$ or $\alpha, \bar{\alpha} = A, B; m \neq \bar{m} = F, S$ and $n = 0, \pm 1, \pm 2 \ldots; l' = \pm 1$) with $r = +1$ for $\to$ and $r = -1$ for $\leftarrow$, respectively. Note that in this approximation scheme the $W_R$, $W_D$ in Eq. (45) follow from the $\tilde{W}_R$, $\tilde{W}_D$ in Eq. (38) by replacing the $\tilde{C}_n^{m\alpha}$ by the $C_n^{m\alpha}$. The K's in Eq. (45) are called elementary rates because these quantities describe the rate of change of the concentrations of species due to specific elementary processes.

According to Eqs. (39), (40), (38), (28) and (43) we obtain for the elementary rates (45) the following detailed balance conditions[22, 23]

$$\frac{K_n^{F\alpha \to S\alpha}}{K_n^{S\alpha \to F\alpha}} = \exp[-(\delta F)_n^{F\alpha \to S\alpha}/K_B T] \qquad (46\,a)$$

and

$$\frac{K_{n+l' \to n}^{m\alpha \to \bar{m}\bar{\alpha}}}{K_{n \to n+l'}^{\bar{m}\bar{\alpha} \to m\alpha}} = \exp[-(\delta F)_{n+l' \to n}^{m\alpha \to \bar{m}\bar{\alpha}}/K_B T] \qquad (46\,b)$$

where

$$(\delta F)_n^{F\alpha \to S\alpha} = \frac{1}{N}\left[\frac{\partial F}{\partial C_n^{S\alpha}} - \frac{\partial F}{\partial C_n^{F\alpha}}\right] = -(\delta F)_n^{S\alpha \to F\alpha} \qquad (47\,a)$$

and

$$(\delta F)_{n+1'\to n}^{m\alpha\to\bar{m}\bar{\alpha}} = \frac{1}{N}\left[\frac{\partial F}{\partial C_n^{m\alpha}} - \frac{\partial F}{\partial C_{n+1'}^{m\alpha}} + \frac{\partial F}{\partial C_n^{\bar{m}\bar{\alpha}}} - \frac{\partial F}{\partial C_{n+1'}^{\bar{m}\bar{\alpha}}}\right] = -(\delta F)_{n\to n+1'}^{m\alpha\to\bar{m}\bar{\alpha}} \qquad (47\,b)$$

($1/N \ll 1$). F is the system free energy in BWA defined by

$$\langle\tilde{F}[\tilde{C}_n^{m\alpha}]\rangle \xrightarrow{\text{BWA}} \tilde{F}[\langle\tilde{C}_n^{m\alpha}\rangle] \equiv F[C_n^{m\alpha}] \qquad (47\,c)$$

using Eqs. (36 a), (39), (43), and neglecting fluctuation effects.

Finally, analogous to Eq. (30) the interfacial growth rate V is given by[19]

$$V = a \sum_{\alpha=A,B} \sum_{n=-\infty}^{\infty} \frac{d}{dt} C_n^{S\alpha}\,. \qquad (48)$$

Now, it can easily be verified that at equilibrium (i.e. the system free energy assumes a minimum) the left-hand sides of Eqs. (47 a) and (47 b) vanish, and using Eqs. (46 a), (46 b) and (44) the growth rate (48) turns to zero as required. Due to the nonlinear character of the K's with respect to the dependence on the concentrations, the rate equations (44) are coupled nonlinear differential-difference equations which can only be solved numerically for a finite number of layers. In the BWA clearly metastable states exist as artifacts (see[21, 23]). These can be excluded formally using instead of the "layer representation" a "continuum approximation (CA)". The interconnection of the BWA with the CA leads to the so-called continuous Bragg-Williams approximation (CBWA)[19, 20, 22, 23]. The results of a crude mathematical treatment of the rate equations (44) in CBWA will be summarized in the following section. Finally, we note that Eqs. (44) to (48) define the rate-equation approach of two-component growth in BWA[19, 20, 22, 23].

### *3.2.4 General Results of the Continuous Bragg-Williams Approximation Under Steady-State Conditions*

A crude analytical treatment of the rate equation (44) at small deviation from equilibrium under steady-state conditions (SSC) in CBWA has been performed in[19, 21–23] using the following relations:

$$n \xrightarrow{\text{CA}} \xi,\ f_n^\alpha(t) \xrightarrow{\text{CA}} f^\alpha(\xi, t) \xrightarrow{\text{CA + SSC}} f^\alpha(\zeta);\ \frac{d}{dt} f^\alpha(\xi,\ t) \to -\frac{V}{a}\frac{d}{d\zeta} f^\alpha(\zeta) \qquad (49\,a)$$

with

$$\zeta = \xi - \frac{V}{a} t \qquad (49\,b)$$

and

$$f_{n\pm 1}^\alpha(t) \xrightarrow{\text{CA + SSC}} f^\alpha(\zeta) \pm \frac{d}{d\zeta} f^\alpha(\zeta) + \frac{1}{2}\frac{d^2}{d\zeta^2} f^\alpha(\zeta) \qquad (49\,c)$$

neglecting higher-order terms in Eq. (49c). $f_n^\alpha$ is an arbitrary function belonging to a component $\alpha$ in the "layer representation" n, $\xi$ and $\zeta$ are dimensionless continuous length variables and V is the steady-state growth rate. Eq. (49c) is appropriate only for atomically rough interfaces where the growth kinetics is governed by the normal growth mode[19–23]. Choosing, for instance, a symmetric interface dynamics, the K's in Eq. (45) can be constructed from Eq. (46) with the help of Eq. (47) (see[22, 23]). Introducing these K's into Eq. (44) and using Eq. (49), a "solitary-like" solution of Eq. (44) in CBWA can be obtained calling to mind Eqs. (34), (43), (41b), and (49b) (see[22, 23]):

$$x_n(t) \to x[C], \; \eta_n(t) \to \eta[C] \tag{50a}$$

$$C_n(t) \to C[\zeta] = [1 + \exp(4\zeta/\Delta)]^{-1} \tag{50b}$$

where $C[\zeta]$ fulfils the SOS boundary conditions $C[\xi \to \infty] = 0$ and $C[\xi \to -\infty] = 1$ ($\Delta$ measures the interface width in units of the lattice constant). Moreover, it can be shown[19, 23] that in the limiting case $\xi \to \pm\infty$ the rate equation (44) converges to the diffusion equation of the fluid and solid bulk phase, as it must be. From Eq. (50) one obtains equilibrium ($V = 0$) and kinetic ($V \neq 0$) phase diagrams (solidus $x[C = 1, T]$, liquidus $\eta[C = 0, T]$), the steady-state growth rate V as a function of the deviation from equilibrium ($\Delta T = T_{eq} - T$ where $T_{eq}$ is the equilibrium temperature of the system), the interface profile $C[\zeta]$, and the concentration profiles $x[\zeta], \eta[\zeta]$ (see[19–23]). As shown in[79] the computed equilibrium properties are in resonable agreement with the results of the pair approximation for two-component systems with $\omega_\eta = 0$ (see Eq. (13)).

The non-equilibrium properties depend on the following entry parameters[23]: kinetic parameters $\tau_\alpha$, $\tau_{FF}$, $\tau_{SS}$ (time scale for the phase transition $(F\alpha) \rightleftarrows (S\alpha)$ and the diffusion process in the fluid F and solid S phase, respectively, due to (Eq. (45)), energy parameters $\Phi^{m\alpha, \bar{m}\bar{\alpha}}$ and thermodynamic parameters $\mathring{\mu}^{m\alpha}$ ($m, \bar{m} = F, S$; $\alpha, \bar{\alpha} = A, B$). Especially, the kinetic coefficient $\mathcal{K}$, defined by

$$\mathcal{K} = V\Big/\frac{\Delta T}{T_{eq}} \quad (\Delta T = T_{eq} - T) \tag{51a}$$

has been computed as a function of the entry parameters in[23] under the simplified assumptions of free miscibility (i.e. $\omega_\eta = 0$ (see (13))) in the fluid and solid phase, complete wetting of the component B (i.e. $\Phi^{FB,FB} = \Phi^{FB,SB}$) and if $\Phi^{SA,BF} = \Phi^{SB,FA} = \Phi^{FA,FB}$. Only for $\tau_{SS} \to \infty$ (no diffusion in the solid) can Eq. (51a) be expressed analytically. It holds[23]

$$\mathcal{K}^{-1} = \mathcal{K}_A^{-1} + \mathcal{K}_B^{-1} + \mathcal{K}_D^{-1} =$$

$$= \left[\frac{\tau_A}{a}\right]\left[F_1(g) + F_2(g)\frac{\Delta_{eq}}{\delta_A}(K_{eq}^A - 1)\eta_{eq}\left(\frac{\tau_{FF}}{\tau_A}\right)\right] \tag{51b}$$

where $F_1$ and $F_2$ are quantities which depend besides $g = \tau_A/\tau_B$ on the energy and thermodynamic parameters. $K_{eq}^A = x[C = 1, T_{eq}]/\eta_{eq}$ is the bulk equilibrium distribution coefficient for the A-component, $\eta_{eq} \equiv \eta[C = 0, T_{eq}]$ the fraction of the component A in the fluid, $\Delta_{eq}$ the equilibrium interface width and $\delta_A = (\Phi^{FA,FA} - \Phi^{SA,FA})/K_B T_A$ the wetting parameter of the component A ($T_A$ is the melting temperature of the pure A-component). $\mathcal{K}_\alpha$ is the kinetic coefficient due to the relaxation kinetics of the component $\alpha$ = A, B, and $\mathcal{K}_D \sim \delta_A/[\Delta_{eq}(K^A_{eq} - 1)\tau_{FF}]$ is the kinetic coefficient resulting from the diffusion kinetics. From Eq. (51 b) follows that, in contrast to phenomenological diffusion theories of crystal growth[35, 67], a stationary growth rate $V \neq 0$ may exist in the absence of convection. This is due to concentration profiles across the interface. For complete wetting ($\delta_A = 0$) (51 b) leads to $V = 0$, i.e. a stationary growth rate does not exist which is in agreement with the phenomenological theory[35]. The same holds for $\Delta_{eq} \to \infty$ (small concentration profile) and for $\tau_{FF} \to \infty$ (no mass transport in the fluid) as required. For $K^A_{eq} = 1$, a stationary growth rate exists independent of $\delta_A$, $\Delta_{eq}$, and $\tau_{FF}$ which is also in agreement with phenomenological theories. Finally, the terms in the bracket of Eq. (51 b) must be positive because otherwise a stable stationary growth rate does not exist.

For a more detailed discussion of the results of two-component crystal growth in CBWA see[19–23].

In the following section selected problems of two-component crystal growth are discussed using the theoretical background described in Sect. 3.

# 4 Selected Problems

## 4.1 Growth of Ordered Binary Crystals

If the interactions between SA-SB particles are stronger than between SA-SA and SB-SB particles, the system tends to prefer A-B pairs and to form an ABAB... superlattice (at sufficiently low temperatures), i.e. is it required $\omega_1 < 0$ (see Eq. (13)). Ordered AB growth has been studied for the cases:

(I) $\Phi^{SA,SA} = \Phi^{SB,SB} = [1/2]\,\Phi^{SA,SB}$ [3, 4, 61, 62]

and

(II) $\Phi^{SA,SA} = \Phi^{SB,SB} = -\Phi^{SA,SB} > 0$ [45]

in the frame of the Kossel (SOS) model. The degree of ordering of the solid phase in case (I) has been described by SRO and LRO parameters, whereas in case (II) only the SRO parameter is considered and discussed as a function of the driving forces (chemical potential differences) of the particles. It is shown that an increase of the driving force leads to a reduction of the correlated deposition of the particles at the interface and to a breakdown of the ordered structure by the formation of domains which have internally still a regular structure but their sizes decrease smoothly with increasing supersaturation (kinetic disordering[3, 4, 45, 61]). However, large $|\Phi^{SA,SB}|$ and small driving forces give rise to the growth of almost ideally ordered AB crystals. This process can be considered as the formation of the solid compound AB via interfacial chemical reactions. A simple analytical approach to this problem is given in[68] where only particle exchange occurring at kink sites is taken into account.

It was observed[2–4, 45, 61, 62] that the ordered ABAB . . . superlattice structure, which is formed at small supersaturations (undercoolings), is destroyed at higher growth rates. This order-disorder transition in the solid phase induced by the growth kinetics was termed "kinetic phase transition" and appears as the cases (I) and (II). Recently, the kinetic phase transition in binary crystals[2–4] was studied by taking into account also the diffusion process in the fluid phase[61] and by considering a rough-stepped SOS-interface with and without restricted jump heights[62]. Furthermore, a non-SOS lattice model (antiferromagnetic spin $-1$ Ising model)[24, 25] for the crystallization of a binary alloy was investigated. This alloy, for a certain range of model parameters, undergoes simultaneously two phase transitions, crystallization (non-magnetic to magnetic) and an order-disorder (antiferromagnetic) transition of the sublattice ordering in the solid phase. Therefore, two order parameters exist, one for the crystal density (describing the fluid-solid transition within the interface) and one for the sublattice ordering in the solid (describing kinetic disordering). If the formation of the interface between an ordered solid phase and a fluid phase exceeds a critical velocity, a disordered solid phase forms in between because the sublattice ordering cannot follow this speed so that the width of the disordered solid phase increases with time (kinetic phase transition) and the ordered solid becomes dynamically unstable. On the basis of these results a deeper understanding of the mechanism of disordering induced by growth kinetics is possible.

Another characteristic aspect of two-component crystal growth is the incorporation of impurities into the fluid-solid interface in connection with interfacial processes. This problem will be discussed in the following subsection in more detail.

## 4.2 Incorporation of Impurities

As a further example of the application of the presented theory we shall treat the problem of incorporation of impurities into growing crystals within the BWA, taking into account the difference between covered (c) and free (f) particles at the interface. For this type of BWA, the system-free energy F is determined by Eqs. (36 b), (41 b) and (47 c). The interplay between interface structure and impurity distribution was already investigated for equilibrium via minimization of the free energy (Eq. (36 b)) with Eqs. (41 b) and (47 c) in[20]. Bearing in mind Eqs. (35) and (41 b) we use, for convenience, the following notations:

$$n \equiv z,\ C_0^\alpha \equiv \eta^\alpha \left(= \overset{f}{\eta}{}_n^\alpha = \overset{c}{\eta}{}_n^\alpha\right),\ \overset{p}{x}{}_n^\alpha \equiv x_p^\alpha(z),\ C_n \equiv C_1(z),\ C_n^\alpha \equiv C_1^\alpha(z);\ p = c, f \tag{52}$$

where it is assumed that the fluid phase ($\overset{p}{\eta}{}_n^\alpha$; $p = c, f$) is spatially homogeneous. Furthermore, we assume that the component B (impurity) is strongly diluted in the solid: $C_1^B(z) \ll 1$, $C_1(z) \simeq C_1^A(z)$ where the component A builds up the s.c. host lattice. The mean concentration of B-particles $x^B(z)$ within the layer z is given according to Eqs. (35), (41 b) and (52) by

$$C_1(z)\, x^B(z) = C_1(z+1)\, x_c^B(z) + (C_1(z) - C_1(z+1))\, x_f^B(z) \tag{53}$$

where $x_c^B$ and $x_f^B$ are respectively the concentrations of covered and free solid B-particles within the solid part of the layer ($x_c^B(z)$, $x_f^B(z) \ll 1$). Analogously to Eq. (44), the incorporation of impurities into growing crystals is then governed by the following BWA rate equations with distinguished covered and free particles at the interface (see also[69]):

$$\frac{d}{dt} C_1(z) = W_{10}^A \Big[ (C_1(z-1) - C_1(z))\, C_0^A - (C_1(z) - C_1(z+1)) \times \times \exp\Big[\Theta_A - \frac{1}{2}\gamma_{AA} - 2\gamma_{AA} C_1(z)\Big]\Big] \tag{54 a}$$

$$x_c^B(z)\,(1-w)\frac{d}{dt} C_1(z+1) + C_1(z+1)\frac{d}{dt} x_f^B = W_{10}^A\, C_0^B \times \times [C_1(z) - C_1(z+1)]\,[x_f^B(z) - w\, x_c^B(z)] \tag{54 b}$$

$$\frac{d}{dt}[C_1(z)\,x^B(z)] = W_{10}^B \Big[ (C_1(z-1) - C_1(z))\, C_0^B - (C_1(z) - C_1(z+1)) \times \times x_f^B \exp\Big(\Theta_B - \frac{1}{2}\gamma_{AB} - 2\gamma_{AB} C_1(z)\Big)\Big] \tag{54 c}$$

where $w = \exp[1/2\,(\gamma_{AA} - \gamma_{AB})]$, $\gamma_{AA} = 2\,\Phi_{AA}/K_BT$, $\gamma_{AB} = 2\,\Phi_{AB}/K_BT$, $\Phi_{\alpha\beta} = \varphi_{01}^{\alpha\beta} - \varphi_{11}^{\alpha\beta}$, $\Theta_\alpha = (3\,\Phi_{\alpha\alpha} - \Delta\mathring{\mu}_\alpha)/K_BT$ with $\alpha, \beta = A, B$. In Eq. (54), for simplicity, we have assumed complete wetting ($\varphi_{10}^{\alpha\beta} = \varphi_{00}^{\alpha\beta}$), free miscibility ($\omega_0 = 0$) and infinitely fast diffusion in the fluid state and neglected interlayer diffusion in the solid. If we ignore the difference between covered and free particles, i.e. $x_c^B = x_f^B \equiv x^B$, Eqs. (54 a) and (54 c) go over to the usual rate equations in BWA (Eq. (29) for $C_1^B(z) \ll 1$). At equilibrium (l.h.s. of Eqs. (54 a) to (54 c) are equal to zero), we obtain from Eq. (54) the results of[20]. Eq. (54 a) describes the growth of the pure A-crystal, and Eqs. (54 b) and (54 c) depict the problem of the incorporation of impurities. As shown in[69], Eqs. (54 a) to (54 c) can be treated analytically in the steady-state continuum approximation (49 a) using the following relations:

$$C_1(\zeta) \sim \begin{cases} (1 - \exp[\bar{r}\,\zeta]) & \text{for } \zeta \to -\infty \\ \exp[-\bar{r}\,\zeta] & \text{for } \zeta \to \infty\,; \end{cases} \quad x_{c(f)}^B(\zeta) = a_{c(f)} + b_{c(f)}\, C_1(\zeta) \tag{55}$$

where $\zeta$ is defined by Eq. (49 b). The free parameters $\bar{r}$, $R = (V/a)\,(W_{10}^A)^{-1}$, $a_{c(f)}$ and $b_{c(f)}$ are determined requiring that Eqs. (54 a), (54 b) and (54 c) are fulfilled for $\zeta \to \pm\infty$, using Eqs. (49 a), (49 b) and (55). For the impurity concentration in the bulk solid $x^B \equiv x_c^B(-\infty) = a_c + b_c$ as a function of the steady-state growth rate R we obtain with Eq. (55) from Eq. (54) (see also[69]):

$$x^B(R) = \mathscr{H} C_0^B \frac{\exp(\bar{r})\,[RQ + UE] + \exp(-\bar{r})[RQ + LM]}{(RQ + LM)[\exp(\bar{r})[RQ + EL] + RQ + UE - \mathscr{H}^2 w\bar{r}E]} = K^B C_0^B \tag{56 a}$$

with

$\mathcal{H} = (e^{\bar{r}} - 1)/\bar{r}$; $Q = W_{10}^A/W_{10}^B$; $U = R/C_0^B$; $L = U(1-w) + \mathcal{H}w$;
$E = \exp[\Theta_B - 5\gamma_{AB}/2] - RQ\bar{r}$; $M = \exp[\Theta_B - \gamma_{AB}/2] + RQ\mathcal{H}\bar{r}$; $\bar{r} = \ln(1/u)$;
$u = 1 - [C_{0,eq}^A e^{\gamma_{AA}} \times (1 - e^{-2\gamma_{AA}})]/[C_0^A + C_{0,eq}^A e^{\gamma_{AA}}]$; $C_0^B = 1 - C_0^A$.

The growth rate R is given by

$$R = [(u-1)/\ln u][(C_0^A - C_{0,eq}^A)(C_0^A + C_{0,eq}^A)]/[C_0^A + e^{-\gamma_{AA}} C_{0,eq}^A] \tag{56b}$$

with $C_{0,eq}^A = \exp[\Theta_A - 3/2\,\gamma_{AA}]$ as equilibrium concentration of the A-component in the fluid. To obtain concrete results the following entry parameters are fixed:

$$\Theta_A = 0.7542; \quad \gamma_{AA} = 2.5; W_{10}^A/W_{10}^B = 18 \tag{57}$$

In contrast, the remaining parameters are varied for four different special cases (see Fig. 3). Figure 3 describes the dependence of the impurity content in the crystal on the growth rate according to Eq. (56a) in comparison with results obtained by the pair approximation (PA), which was shown to be in quantitative agreement with the Monte Carlo simulations (see[69] and Sect. 3.1). In spite of the crude approximation (55) both results agree surprisingly good. Moreover, also the dependence of the growth rate (56b) on the concentration $C_0^A$ is in good agreement with the pair approximation[69]. For small growth rates, the impurity concentration is the higher, the larger $[\gamma_{AA} - \gamma_{AB}]$. Ignoring the difference between covered and free particles, i.e. starting from Eq. (29), the same qualitative features as in Fig. 3 were found in[69] to[71]. However, it can be shown that this neglection underestimates the increase of the impurity content of component B with the growth rate (see Fig. 3). According to Chernov's formula (Eq. (5.7)) in[3] the impurity

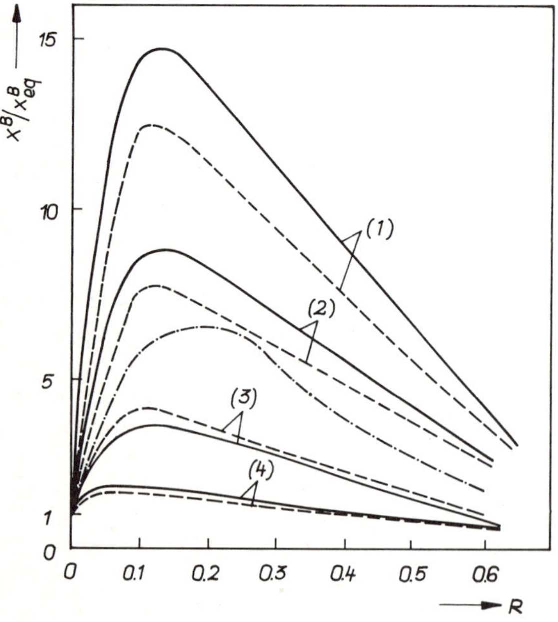


**Fig. 3.** Dependence of the reduced impurity concentration $x^B/x^B_{eq}$ on the dimensionless growth rate R according to pair approximation (PA) (solid lines) and according to Eq. (56a) in improved BWA (with covered and free surface sites) (broken lines): $x^B_{eq} = C_0^B K^B_{eq}$ where $K^B_{eq} = \exp[-3/2\,\gamma_{AA} + 3\gamma_{AB} - \Theta_B]$ is the equilibrium distribution coefficient for the impurity B. The numbers at the lines correspond to the cases: (1) $\Theta_B = 2.1$; $\gamma_{AB} = 0$ (2) $\Theta_B = 3.355$; $\gamma_{AB} = 0.5$ (3) $\Theta_B = 5.855$; $\gamma_{AB} = 1.5$ (4) $\Theta_B = 8.355$; $\gamma_{AB} = 2.5$ where the fixed remaining entry parameters are given in Eq. (57). The dash-pointed line is a numerical solution of the BWA equation (29) corresponding to case (1)

concentration in the crystal may also either increase or decrease with the growth rate as a function of the binding energy of the impurity in the solid. However, the increase of $x^B$ with R is much lower than shown in Fig. 3. Therefore, the capture of impurities is intimately related to the details of the surface sites and the crystal growth process[27, 72].

## 5 Conclusions and Future Trends

At present, the master-equation approach in the case of micro- or macrovariables is the starting point for a unified microscopic theory of the growth of one- and multicomponent systems. In the frame of a multilayered SOS-lattice model, the equilibrium and non-equilibrium properties of a multicomponent system can be described either analytically by different statistical approximation schemes (SPA, PA, BWA) or by the direct Monte Carlo computer simulation of the master equation. The results have shown that the pair approximation (PA) agrees satisfactorily well with the Monte Carlo results. Moreover, special growth features of binary Kossel crystals can be described qualitatively by an improved BWA (with covered and free surface sites). To include cluster effects beyond the pair approximation the master-equation approach must be formulated in terms of cluster variables to describe the state of the interface at nucleation (see[26]). Although basic features of the growth characteristics of multicomponent systems can be understood qualitatively within the frame of the SOS lattice model, modelling of the fluid state (especially the melt) in terms of the lattice model represents a crucial problem (see[18]). Crystal growth takes place at the interface between the crystal itself and its environment. Growth from vapour (gas) or from solution phase can be described as "dilute environment problem" (where the concentration of the fluid phase is very low in comparison with the growing solid phase). In contrast, the melt growth must be considered as "dense environment problem" since the liquid phase usually has almost the same atomic density as the solid phase and, in addition, it is believed that both phases exhibit a very similar short-range order with respect to the atomic structure. Therefore, when modelling the solid-liquid (melt) interface we have to take into account the atomic structure of the environment explicitly, in contrast to the case of "dilute environment". Thus, a more realistic description of melt growth requires a molecular theory of the solid-liquid interface (see[51]). Furthermore, also a molecular theory of fluid-fluid interfaces[77] is useful for the understanding of nonuniform fluid systems (see[78]).

Finally, this article deals mainly with the representation of the basic features of a unified microscopic theory of the growth of two-component systems. A survey of special problems generally discussed here is given in Table 1. In future, topics like the investigation of lateral growth modes (step and spiral growth), formation of chemical compounds, nucleation and growth of clusters as well as the study of the influence of interface kinetics on the interface stability in multicomponent systems will be important tasks.

## 6 Nomenclature

List of frequently used symbols

| | |
|---|---|
| $C_j^{m\alpha}$ | occupation number of the lattice site j with a particle $m\alpha$ belonging to the phase m = F, S and component $\alpha$ = A, B as microvariable, see Eq. (1) |
| $\Phi^{m\alpha,\bar{m}\bar{\alpha}} \equiv \varphi_{\eta\bar{\eta}}^{\alpha\bar{\alpha}}$ | nearest neighbour interaction (bond) energy between particles $m\alpha$ and $\bar{m}\bar{\alpha}$ ($m, \bar{m}$ = F, S; $\alpha, \bar{\alpha}$ = A, B) or $\alpha\eta$ and $\bar{\alpha}\bar{\eta}$ ($\eta, \bar{\eta}$ = 0.1), see Eq. (3c) |
| $m, \eta$ | as subscript: refers to a fluid if m = F or $\eta = 0$ and to a solid if m = S or $\eta = 1$ |
| $\mathring{\mu}^{m\alpha} \equiv \mathring{\mu}_\eta^\alpha$ | standard chemical potential of particle $m\alpha$ or $\alpha\eta$ |
| $\Theta_\eta^\alpha$ | entropy factor, see Eq. (3d) |
| $C_j^S \equiv \eta_j$ | occupation number of the lattice site j with solid particles as microvariable, see Eq. (3a) |
| $\varrho(g, t)$ | probability that the system is in a microstate g at time t |
| $W(g \rightleftarrows g')$ | conditional (transition) probability per unit time for the transition $g \leftrightarrows g'$, see Eqs. (6) to (12) |
| g | a microstate, see Eq. (3b) |
| $\hat{H}(g)$ | Hamiltonian in the microstate g |
| $\omega_\eta$ | mixing energy, see Eq. (13) |
| $\varrho^{(l)}$ | distribution function of order l = 1, 2, ..., see Eqs. (16), (17) |
| $n \equiv z$ | layer index ($n = 0, \pm 1, \pm 2, \ldots$) |
| $\langle D \rangle$ | expectation value of the quantity D, see Eqs. (21), (41a) |
| $\langle C_j^{m\nu} \rangle = C_\eta^\nu(n) \equiv C_n^{m\nu}$ | macroscopic (averaged) concentration of a $\nu$-particle in the state m or $\eta$ on lattice sites of the layer n, see Eqs. (23), (43) |
| $C_{\eta\eta'}^{\nu\nu'}(n)$ | macroscopic (averaged) concentration of pairs of neighbouring particles $(\nu, \eta)$ and $(\nu', \eta')$ on the sites of the n-th layer, see Eq. (24a) |
| $C_{\eta\eta'}^{\nu\nu'}(n, n-1)$ | macroscopic (averaged) concentration of pairs of neighbouring particles $(\nu, \eta)$ and $(\nu', \eta')$ where $(\nu, \eta)$ belongs to layer n and $(\nu', \eta')$ to layer n − 1, see Eq. (24b) |
| $\Phi_{\nu\gamma}$ | bonding difference $(\varphi_{01}^{\nu\gamma} - \varphi_{11}^{\nu\gamma})$ ($\nu, \gamma$ = A, B) |
| $\Theta_\nu$ | difference between entropy factors $\Theta_1^\nu - \Theta_0^\nu$ ($\nu$ = A, B), see Eq. (3d) |
| $W_{10}^\nu$ | attachment probability of particle $\nu$ passing from state $\eta = 0$ (or m = F) to $\eta = 1$ (or m = S), see Eq. (9) |
| R | dimensionless interfacial growth rate, see Eq. (30) |
| V | interfacial growth rate, see Eq. (48) |
| $\tilde{C}_n^{m\alpha}$ | concentration (normalized intensive macrovariable) of an $\alpha$-particle in the state m within the layer n (m = F, S; $\alpha$ = A, B), see Eqs. (33), (34) |
| T | absolute temperature (of the heat bath) |
| $\tilde{F}[\tilde{C}_n^{m\alpha}]$ | Helmholtz free energy representation in macrovariables $\tilde{C}_n^{m\alpha}$, see Eq. (36a) |
| $F[C_n^{m\alpha}]$ | Helmholtz free energy representation in macroscopic (averaged) concentrations $C_n^{m\alpha}$, see Eq. (47c) |
| $P[\tilde{y}; t]$ | probability of finding the system at time t in a macrostate with the macrovariables $\tilde{y}$, see Eq. (37) |
| $W(\tilde{y} \to \tilde{y}')$ | conditional (transition) probability per unit time for the transition $\tilde{y} \to \tilde{y}'$, see Eq. (38) |

| | |
|---|---|
| $\tilde{Y}$, $\tilde{y}$ | extensive and normalized intensive macrovariable, respectively |
| $\tilde{x}_n^\alpha$, $\tilde{x}_n^{p\alpha}$ | fraction of the $\alpha$-component in the solid phase within the layer n (p = c, f) as normalized intensive macrovariable |
| $\tilde{\eta}_n^\alpha$, $\tilde{\eta}_n^{p\alpha}$ | fraction of the $\alpha$-component in the fluid phase within the layer n (p = c, f) as normalized intensive macrovariable |
| $\tilde{C}_n$ | concentration of the solid in the layer n as normalized intensive macrovariable |
| $K_n^{F\alpha \rightleftarrows S\alpha}$ | elementary rates of the processes $F\alpha \rightleftarrows S\alpha$ within the layer n, see Eq. (45 a) |
| $K_{n+l' \rightleftarrows n}^{m\alpha \rightleftarrows \bar{m}\bar{\alpha}}$ | elementary rates of the diffusion processes of species $m\alpha$ and $\bar{m}\bar{\alpha}$ migrating from layer n + l′ to layer n and vice versa, see Eq. (45 b) |
| $(\delta F)_n^{F\alpha \rightleftarrows S\alpha}$ | change of the system free energy due to the processes $F\alpha \rightleftarrows S\alpha$ occuring within the layer n, see Eq. (47 a) |
| $(\delta F)_{n+l' \rightleftarrows n}^{m\alpha \rightleftarrows \bar{m}\bar{\alpha}}$ | change of the system free energy due to the diffusion of species $m\alpha$ and $\bar{m}\bar{\alpha}$ from layer n + l′ to layer n and vice versa, see Eq. (47 b) |
| $\mathcal{K}$ | kinetic coefficient, see Eq. (51) |
| $x_n^\alpha$, $x_n^{p\alpha} \equiv x_p^\alpha(n)$ | macroscopic (averaged) fraction of the $\alpha$-component in the solid phase within the layer n, see Eqs. (41 b) and (52) (p = c, f) |
| $\eta_n^\alpha$, $\eta_n^{p\alpha} \equiv \eta_p^\alpha(n)$ | macroscopic (averaged) fraction of the $\alpha$-component in the fluid phase within the layer n, see Eqs. (41 b) and (52) (p = c, f) |
| $C_n \equiv C_1(n)$ | macroscopic (averaged) concentration of the solid in the layer n, see Eqs. (41 b) and (52) |
| $K_{eq}^\alpha(K^\alpha)$ | equilibrium (nonequilibrium) distribution coefficient of the component $\alpha$; $K^\alpha = x^\alpha \mid C_0^\alpha$ ($\alpha$ = A, B) |
| $x^B$ | concentration of the impurity B in the bulk solid, see Eq. (56 a) |
| $W_{B0}^{A0}$ | diffusion probability of particles A and B in the fluid phase $\eta = 0$, see Eq. (14). |

# 7 References


1. As an introduction into problems of crystal growth, the reader is referred to:
   a) Elwell, D., Scheel, H. J.: Crystal growth from high-temperature solutions, London, Academic Press 1975
   b) Parker, R. L.: Crystal growth mechanisms: energetics, kinetics and transport, in: Solid-State Physics (eds.) Seitz, F., Turnbull, D., Ehrenreich, H., vol. 25, p. 151, New York–London, Academic Press 1970
   c) Mullin, J. W. (ed.): Industrial crystallization, New York, Plenum Press 1976
   d) Freyhardt, H. C. (ed.): Crystals/growth, properties, and applications, vol. *1*, (1978), vol. *2*, (1979), vol. *3*, (1980), Berlin – Heidelberg – New York, Springer-Verlag
2. Chernov, A. A.: Int. Conf. on Crystal Growth (ed.) Peiser, H. S., p. 25, Oxford, Pergamon 1967, see: J. Phys. Chem. Solids, Suppl. *1*, 25 (1967)
3. Chernov, A. A.: Sov. Phys.-Uspekhi *13*, 1011 (1970)
4. Chernov, A. A., Lewis, J.: J. Phys. Chem. Sol. *28*, 2185 (1967)
5. Temkin, D. E.: a) J. Cryst. Growth *5*, 193 (1969); Sov. Phys.: b) Kristallografiya *14*, 423 (1969); c) Kristallografiya *15*, 421 (1970); *15*, 428 (1970) and *19*, 467 (1974)

6. Temkin, D. E.: Sov. Phys.: Rost Kristallov (ed.) Chernov, A. A., vol. 11, p. 327, Yerevan, Nauka 1975
7. Cherepanova, T. A.: Sov. Phys: J. Appl. Mechn. and Techn. Phys. *5*, 120 (1978); *6*, 96 (1978)
8. Cherepanova, T. A.: Dokl. Akad. Nauk. SSSR *238*, 162 (1978)
9. Cherepanova, T. A.: Phys. Stat. Sol. (a) *58*, 469 (1980); *59*, 371 (1980)
10. Cherepanova, T. A., Didrihsons, G. T.: Phys. Stat. Sol. (a) *59*, 633 (1980)
11. Cherepanova, T. A., Kiselev, V. F.: Kristall and Technik *14*, 545 (1979)
12. Cherepanova, T. A., Didrihsons, G. T.: Kristall und Technik *14*, 1501 (1979)
13. Cherepanova, T. A.: Sov. Phys. in: Proc. of 11th Int. Mineral. Association Meet. Crystal Growth, p. 145, Moscow, Izd. Nauka 1980
14. Cherepanova, T. A.: Cryst. Res. Technol. *16*, 307 (1980)
15. Cherepanova, T. A., Dzelme, J. B.: Cryst. Res. Technol. *16*, 399 (1981)
16. Cherepanova, T. A., Shirin, A. V., Borisov, V. T.: Computer simulation of crystal growth from solution, in: Industrial Crystallization (ed.) Mullin, J. W., p. 113, New York, Plenum Press 1976
17. Cherepanova, T. A., Kiselev, V. F.: Sov. Phys.: Kristallografiya *24*, 327 (1979); *25*, 1010 (1980)
18. Cherepanova, T. A.: J. Cryst. Growth *52*, 319 (1981)
19. Temkin, D. E.: Sov. Phys.: Kristallografiya *23*, 1151 (1978)
20. Temkin, D. E.: Sov. Phys.: Kristallografiya *24*, 421 (1979)
21. Temkin, D. E.: J. Cryst. Growth *52*, 299 (1981)
22. Pfeiffer, H.: Phys. Stat. Sol. (b) *93*, K 149 (1979)
23. Pfeiffer, H., Haubenreisser, W.: Phys. Stat. Sol. (b) *96*, 287 (1979) and *101*, 253 (1980)
24. Saito, Y.: J. Chem. Phys. *74*, 713 (1981)
25. Saito, Y., Müller-Krumbhaar, H.: J. Chem. Phys. *74*, 721 (1981)
26. Müller-Krumbhaar, H.: Kinetics of crystal growth, in: Current Topics in Materials Science (ed.) Kaldis, E. vol. 1, Chap. 1, Amsterdam, North-Holland Publ. Comp. 1978
27. Gilmer, G. H., Jackson, K. A.: Computer simulation of crystal growth, in: Current Topics in Materials Science (ed.) Kaldis, E., Scheel, H. J., vol. 1, Chap. 1.2, p. 80, Amsterdam, North-Holland Publ. Comp. 1977
28. Temkin, D. E.: Crystallization processes, p. 15, New York, Consultants Bureau 1966
29. Leamy, H. J., Gilmer, G. H., Jackson, K. A.: Statistical thermodynamics of clean surfaces, in: Surface Physics of Materials (ed.) Blakely, J. M., vol. 1, p. 121, New York, Academic Press 1975
30. Bennema, P., Gilmer, G.: Crystal growth: a introduction (ed.) Hartman, P., p. 263, Amsterdam, North-Holland Publ. Comp. 1973
31. Van Leeuwen, C.: The lattice gas interface, Thesis, Delft 1977
32. Burton, W. K., Cabrera, N., Frank, F. C.: Phil. Trans. Roy. Soc. (London) A *243*, 299 (1951)
33. Gilmer, G. H.: J. Cryst. Growth *35*, 15 (1976)
34. Welberry, T. R., Miller, G. H., Pickard, D. K.: Proc. Roy. Soc. (London) A *367*, 175 (1979)
35. Brice, J. C.: The growth of crystals from liquids, Amsterdam, North-Holland Publ. Comp. 1973
36. Ohara, M., Reid, R. C.: Modelling crystal growth rates from solution, New York, Englewood Cliffs 1973
37. Haken, H.: Rev. Mod. Phys. *47*, 67 (1975)
38. Müller-Krumbhaar, H.: Monte Carlo simulation of crystal growth, in: Topics in Current Physics (ed.) Binder, K. vol. 7, Chapt. 7, p. 261, Berlin, Springer-Verlag 1979
39. Van der Eerden, J. P., Bennema, P., Cherepanova, T. A.: Progr. Cryst. Growth Charact., vol. *1*, p. 219, London, Pergamon Press 1978
40. Kawasaki, K.: Kinetics of Ising models, in: Phase Transitions and Critical Phenomena (ed.) Domb, C., Green, M. S., vol. 3, p. 443, New York, Akademic Press 1972
41. Kitahara, H., Ross, J.: Statistical mechanical theory of the kinetics of phase transitions, in: Fluctuation Phenomena (ed.) Montroll, E. W., Lebowitz, J. L., p. 229, Amsterdam, North-Holland Publ. Comp. 1979
42. Müller-Krumbhaar, H.: Z. Phys. B *25*, 287 (1976)
43. Kubo, R., Matsuo, K., Kitahara, K.: J. Stat. Phys. *9*, 51 (1973)
44. Saito, Y., Müller-Krumbhaar, H.: J. Chem. Phys. *70*, 1078 (1979)
45. Cherepanova, T. A., Van der Eerden, J. P., Bennema , P.: J. Cryst. Growth *44*, 537 (1978)
46. Weeks, J. D., Gilmer, G. H.: J. Cryst. Growth *33*, 21 (1976)

47. Van der Eerden, J. P. et al.: J. Appl. Phys. *48*, 2124 (1977)
48. De Fontaine, D.: Configurational thermodynamics of solid solutions, in: Solid-State Physics (ed.) Ehrenreich, H., Seitz, F., Turnbull, D., vol. 34, p. 73, New York, Academic Press 1979
49. Kikuchi, R.: J. de Physique Coll. C7, suppl. 12, *38*, C7–307 (1977)
50. Neumann, H. P.: Phys. Rev. B *14*, 4146 (1976)
51. Haymet, A. D. J., Oxtoby, D. W.: J. Chem. Phys. *74*, 2559 (1981)
52. Müller-Krumbhaar, H.: Phys. Rev. B *10*, 1308 (1974)
53. Pfeiffer, H., Haubenreisser, W., Klupsch, Th.: Phys. Stat. Sol. (b) *83*, 129 (1977)
54. Müller-Krumbhaar, H., Binder, K.: J. Stat. Phys. *8*, 1 (1973)
55. Stanley, H. E.: Introduction to Phase Transitions and Critical Phenomena, Oxford, Clarendon Press 1971
56. Binder, K.: Phys. Rev. B *8*, 3423 (1973)
57. Mathews, P. M., Shapiro, I. I., Falkoff, D. I.: Phys. Rev. *120*, 1 (1960)
58. De Groot, S. R., Mazur, P.: Nonequilibrium thermodynamics, Amsterdam, North-Holland Publ. Comp. 1962
59. Pfeiffer, H.: Phys. Stat. Sol. (b) *101*, K117 (1980)
60. Kikuchi, R.: Phys. Rev. B *22*, 3784 (1980)
61. Temkin, D. E.: Sov. Phys.: Kristallografiya *23*, 650 (1978); *15*, 884 (1970)
62. Baikov, Yu. A. et al.: Phys. Stat. Sol. (a) *61*, 435 (1980)
63. Burley, D. M.: Closed form approximations for lattice systems, in: Phase Transitions and Critical Phenomena (ed.) Domb, C., Green, M. S., vol. 3, p. 329, New York, Academic Press 1972
64. Langer, J. S., Sekerka, R. F.: Acta Met. *23*, 1225 (1975)
65. Chan, Sai-Kit.: J. Chem. Phys. *67*, 5755 (1977)
66. Van Kampen, N. G.: Adv. Chem. Phys. *34*, 245 (1976)
67. Müller-Krumbhaar, H.: J. Chem. Phys. *63*, 5131 (1975)
68. Pfeiffer, H.: Phys. Sol. Stat. (a) *65*, 637 (1981)
69. Didrihsons, G. T., Pfeiffer, H.: Phys. Stat. Sol. (a) *71*, 169 (1982)
70. Pfeiffer, H.: Phys. Stat. Sol. (b) *99*, 139 (1980)
71. Pfeiffer, H.: J. Cryst. Growth *52*, 350 (1981)
72. Thurmond, C. D., in: Semiconductors (ed.) Hannay, N. B., chapt. 4, New York, Reinhold 1959
73. Woodruff, D. P.: The solid-liquid interface, London, Cambridge University, Press 1973
74. Nason, D. O.: The structure of the solid-liquid interface, Thesis, Stanford University 1971
75. Broughton, J. Q., Bonissent, A., Abraham, F. F.: J. Chem. Phys. *74*, 4029 (1981)
76. Cape, J. N., Woodcock, L. V.: J. Chem. Phys. *73*, 2420 (1980)
77. Bongiorno, V., Scriven, L. E., Davis, H. T.: J. Colloid Interface Sci. *57*, 462 (1976)
78. Abraham, F. F.: Phys. Rep. (Rev. Sect. Phys. Lett.) *53*, 93 (1979)
79. Didrihsons, G. T., Pfeiffer, H.: Phys. Stat. (a) *71*, K79 (1982)

# Statistics of Surfaces, Steps and Two-Dimensional Nuclei: A Macroscopic Approach

**V. V. Voronkov**

Institute of Rare Metals, B. Tolmachevskiy 5, 109017 Moscow, USSR

*A general theory of equilibrium and kinetic properties of clean crystal surfaces can be developed if a surface is described by positions of its small but macroscopic elementary portions rather than by detailed distribution of surface atoms. Thus the concept of growth units is not used; the properties of singular and vicinal surfaces are directly expressed through the specific free energy and the kinetic coefficient of steps.*



Crystals 9

# 1 List of Symbols

| | |
|---|---|
| $A_e$ | coefficient in step free energy expansion in powers of a slope p |
| a | atomic spacing |
| $C_0(\xi)$ | equilibrium nucleus concentration per unit interval of size variable $\xi$ |
| D, $D_i$ | diffusivity of a step element |
| $D(\xi)$ | diffusivity corresponding to random walk of a nucleus size |
| $D_e$ | effective diffusivity defined by expr. (60) |
| $\hat{D}$ and $\hat{D}'$ | diffusion matrices for variables $\eta_i$ and $\eta_i'$, respectively |
| $F_{cr}$ | work of critical nucleus formation |
| f | growth driving force defined by exprs. (5), (6) |
| g | number of atoms in a nucleus |
| H | heat of fusion (per atom) |
| h | step height (i.e. the distance between adjacent close-packed atomic planes) |
| $h_e$ | kink depth (i.e. the distance between adjacent close-packed atomic rows) |
| I | nucleation rate |
| $k_0$ | Boltzmann's constant |
| L | step length along the x-axis |
| $\Delta l$, $\Delta l_i$ | step element length |
| $l_c$ | surface correlation length |
| M | number of equivalent steps of the lowest kink density |
| **m** | unit normal to a step |
| $\mathbf{m}_0$ | unit normal to close-packed atomic rows |
| $N_c$ | number of atoms in the solid phase |
| n | number of step elements constituting a nucleus boundary |
| **n** | unit normal to a surface |
| $\mathbf{n}_0$ | unit normal to close-packed atomic planes |
| $p = \tan\varphi$ | slope of a step |
| $p_\lambda = \tan\Theta_\lambda$ $(\lambda = 1, 2)$ | slopes of a surface |
| $Q_{ik}$ | correlation function for step elements i and k |
| q | number of atoms per unit area of a monoatomic layer |
| $q_c$ | atomic density in the crystal |
| $q_m$ | atomic density in the melt |
| $\hat{R}$ and $\hat{R}'$ | rate matrices for original variables $\eta_i$ and new variables $\eta_i'$, respectively |
| $R_i$ and $R_i'$ | eigenvalues of matrices $\hat{R}$ and $\hat{R}'$, respectively |

| | |
|---|---|
| $r_{cr}$ | critical nucleus radius |
| $r_n$ | nucleation zone size |
| S | area of surface projected onto the $x_1 x_2$-plane |
| $\Delta s$ | surface element area |
| T | absolute temperature |
| $T_0$ | melting point |
| $\Delta T$ | supercooling |
| $\Delta T_{max}$ | maximum supercooling for a non-isothermal facet |
| t | time |
| $t_e$ | parameter having the formal meaning of diffusion time |
| $\hat{U}$ | variable transformation matrix |
| V | normal growth rate |
| v | normal velocity of a step |
| $v_\infty$ | normal velocity of a straight step |
| $v_i$ | mean velocity of the i-th step element at given element positions |
| $y_i$, y (x) | step element position |
| Z | partition function |
| $dZ_s$ | surface contribution to Z at given element positions |
| $dZ_e$ | step contribution to Z at given element positions |
| $z_i$, $z(x_1, x_2)$ | surface element position |
| $\alpha$ | specific free energy of a step |
| $\alpha^*$ | reduced value of $\alpha$ |
| $\alpha_e$ | stability coefficient of a step |
| $\alpha_e^*$ | reduced value of $\alpha_e$ |
| $\beta$ | kinetic coefficient of a step |
| $\beta^*$ | reduced value of $\beta$ |
| $\beta_0$ | kinetic coefficient of a step with minimum kink density |
| $\zeta$ | curvature of a step |
| $\zeta_1$, $\zeta_2$ | principal curvatures of a surface |
| $\eta_i$, $\eta(\varphi)$ | normal displacement of nucleus boundary element |
| $\Theta$ | tilt angle of a vicinal surface |
| $\lambda$ | parameter defined by expr. (96) |
| $\nu$ | number of elements for a step segment |
| $\xi$ | variable describing a nucleus size |
| $\varrho$ | kink density |
| $\varrho_0$ | minimum value of $\varrho$ |
| $\sigma$ | specific free energy of a surface |
| $\sigma^*$ | reduced value of $\sigma$ |
| $\sigma_{\lambda\mu}$ | stability tensor of a surface |
| $\sigma_{\lambda\mu}^*$ | reduced stability tensor |
| $\sigma_1$, $\sigma_2$ | principal values of $\sigma_{\lambda\mu}$ |
| $\tau$ | time lag of nucleation |
| $\tau_f$ | fluctuation time of a step |
| $\tau_v$ | drift time of a step |

| | |
|---|---|
| $\tau_d$ | randomization time for a vicinal surface |
| $\hat{\phi}$ and $\hat{\phi}'$ | free energy matrices for variables $\eta_i$ and $\eta_i'$, respectively |
| $\phi_i$ and $\phi_i'$ | eigenvalues of $\hat{\phi}$ and $\hat{\phi}'$, respectively |
| $\phi$ | absolute value of $\phi_3'$ |
| $\varphi$ | tilt angle of a step |
| $\varphi_i$ | tilt angle of the i-th element of reference critical contour |
| $\chi$ | coefficient characterizing diffusivity of a step |
| $\Psi$ | free energy for a given state of either a surface (Sect. 2) or a step (Sect. 3 and 4) |
| $\Psi_{ef}$ | effective free energy determining the position distribution for a surface (a step) at $f \neq 0$ |
| $\psi$, $\psi_i$ | given element contribution to $\Psi$ |
| $\omega$ | kinetic factor in nucleation rate |

## 2 Introduction

To analyze the processes on a surface of a growing crystal one commonly uses the concept of attachment and detachment of individual atoms (or growth units). However, this model is too idealized in some cases (e.g. growth from the melt). Besides, it is unnecessarily detailed if the characteristic surface scale (e.g. critical radius of a two-dimensional nucleus) is much greater than the atomic spacing. It is possible to develop a more general theory that deals with displacements of small portions of interface rather than with individual surface atoms. In particular, the growth processes on singular and vicinal surfaces are reduced to tangential motion of steps; this process is again treated as displacements of small portions of a step. This approach is especially useful for a crystal-melt interface. Further we shall confine ourselves to the boundary between one-component crystal and its own melt, although most of the results are of quite general character.

## 3 Statistical Thermodynamics of Crystal Surfaces

### 3.1 Description of Interface States

The crystal bulk can be regarded as a system of parallel atomic planes $\mathcal{P}_j$; the unit normal to these planes is denoted by $\mathbf{n}_0$. The unit normal $\mathbf{n}$ to an interface is assumed to be close to $\mathbf{n}_0$. Let us distinguish a small portion of the interface inside a cylinder of height w and base area $\Delta s$ (Fig. 1). The cylinder base is fixed with respect to one of the $\mathcal{P}_j$-planes; the size w is larger than the width of transient region between the two phases. The total number of atoms inside the cylinder, $N_t$, can serve as a measure of advance of the interface element along the z-axis (which is parallel to $\mathbf{n}_0$). For instance, if the atomic density of crystal, $q_c$, exceeds that of melt, $q_m$, then the increase in $N_t$ corresponds to crystallization. It is convenient to define position z of the interface element according to Gibbs' rule[1])

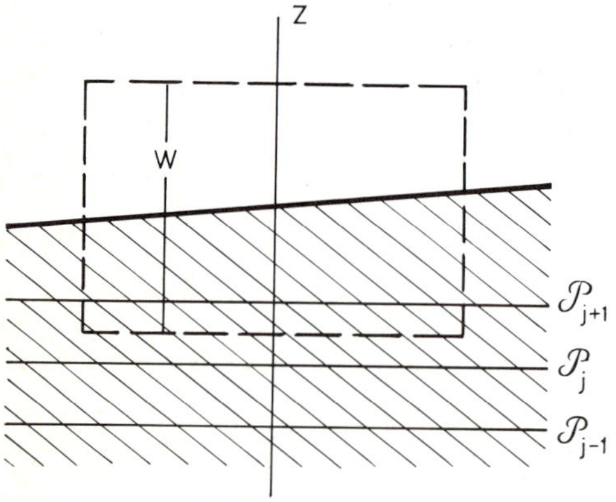


**Fig. 1.** An element of an interface

$$N_t = q_m w \Delta s + (q_c - q_m)(z - z_b) \Delta s \quad (1)$$

where $z_b$ is the coordinate of the cylinder base.

The quantity $N_t$ changes not only due to crystallization (melting), but also to fluctuations in the number of atoms inside a melt element (of volume $\sim w \Delta s$); the fluctuation amplitude $\delta N_t$ is equal[2] to

$$\delta N_t \sim A \sqrt{w \Delta s} \quad (2)$$

The factor A is equal to $q_m (k_0 T K_m)^{1/2}$, where $k_0$ is Boltzmann's constant, T is the absolute temperature, $K_m$ is the isothermal compressibility of melt. It follows from (1), (2) that the element position z has an uncertainty

$$\delta z \sim \frac{A}{|q_c - q_m|} \left( \frac{w}{\Delta s} \right)^{1/2} \quad (3)$$

If $\delta z$ is small in comparison with the distance h between neighboring $\mathscr{P}_j$-planes, then quantity z is suitable for describing infinitesimal displacements of the interface element. The inequality $\delta z \ll h$ is fulfilled at a sufficiently high ratio $\Delta s/w$. For instance, the parameters of silicon[3] correspond to the condition $\Delta s/w \gg 1$ Å which implies that the element size may be rather small.

If a reference cylinder (Fig. 1) were shifted into an equivalent position (i.e. by $\pm h$ along the z-axis) then $z - z_b$ would change by $\mp h$. Consequently, two different positions, z and z′ (with a fixed cylinder) are equivalent to each other if $z' - z = \pm h$. Thus, all the properties of the interface element are periodic in z, the period being equal to plane spacing h.

Further on we shall assume that the crystal and the melt are placed in a cylindrical container (Fig. 2) with a cross-section S; the temperature, volume and total number of atoms are fixed. The interface consists of $\nu$ elements. A state of the interface is described by the set of element positions $z_i$ ($i = 1, \ldots, \nu$). The effective number of atoms in the solid phase is equal to

$$N_c = q_c \Delta s \sum_{i=1}^{\nu} z_i + \text{const} \quad (4)$$

The two-phase system (Fig. 2) may be formed from the initial liquid state. The corresponding change in free energy consists of the bulk term, $-f N_c$ (where f is the decrease in free energy per crystallized atom) and the interface term. The quantity f has the sense of a "driving force of crystallization" and equals

$$f = (\mu_m - p_m \Omega_c) - f_c \quad (5)$$

where $\mu_m$ and $p_m$ are the chemical potential and the pressure of the melt, whereas $f_c$ and $\Omega_c$ are the free energy and the volume (per atom) of the crystal. In particular, if the crystal stresses are simply uniform compression, then f coincides with the chemical potential difference between the two phases and equals[2]

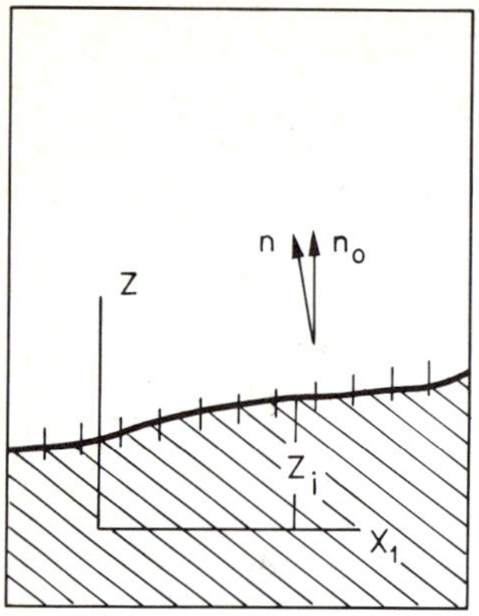


**Fig. 2.** An interface dividing the two phases, placed in a cylindrical container

$$f = \frac{H \cdot (T_0 - T)}{T_0} \tag{6}$$

where $T_0$ is the melting point and H is the heat of fusion (per atom). A correction to Eq. (6) due to stress effect on $f_c$ and $\Omega_c$ is usually negligible (though it becomes essential in some special cases e.g. near the dislocation core[4, 5]).

The partition function of the two-phase system (Fig. 2) with the element positions confined between $z_i$ and $z_i + dz_i$ involves two factors that depend on $z_i$: the bulk factor

$$Z_v = \exp\left(\frac{f N_c}{k_0 T}\right) \tag{7}$$

and the surface factor which is proportional to intervals $dz_i$

$$dZ_s = \exp\left[-\frac{1}{k_0 T} \Psi(z_1, \ldots, z_\nu)\right] \frac{dz_1}{a} \cdots \frac{dz_\nu}{a} \tag{8}$$

where $\Psi(z_1, \ldots, z_\nu)$ is the interface free energy for a given set of positions $z_i$. Each interval $dz_i$ is divided by the normalizing constant a (e.g. atomic spacing) so that the partition function density $\exp(-\Psi/k_0 T)$ is dimensionless. The quantity $\Psi$ includes periodic contributions $\psi(z_i)$ of all the elements and the interaction term $\Psi_{int}$ (which depends on the position differences between neighboring elements)

$$\Psi = \sum_{i=1}^{\nu} \psi(z_i) + \Psi_{int} \tag{9}$$

The complete interface free energy $F_s$ (at $f = 0$) is equal to $-k_0 T \ln Z_s$ where the partition function $Z_s$ is obtained by integrating Eq. (8) over all positions $z_i$ except for the peripheral (adjacent to the container walls) elements. The latter elements should be fixed in a given plane; the normal **n** to this plane defines the interface orientation (Fig. 2). The fixation plane equation can be written in the form

$$z = p_1 x_1 + p_2 x_2 + \text{const} \tag{10}$$

where the coordinates $x_1$, $x_2$ correspond to Cartesian axes normal to the z-axis, while $p_1$ and $p_2$ are the interface slopes in the $zx_1$- and $zx_2$-sections, respectively. The slopes $p_1$, $p_2$ will be used, side by side with **n**, to define the interface orientation.

The interface size can be characterized either by the total interface area $S_t$ or by the area S of its projection onto the $x_1 x_2$-plane ($S = S_t \cos\Theta$ where $\Theta$ is the tilt angle, that is, the angle between **n** and $\mathbf{n}_0$). The interface free energy $F_s$ is proportional to $S_t$ (as well as to S) and equals $\sigma S_t = \sigma^* S$, where $\sigma$ is the specific free surface energy and $\sigma^*$ is its reduced value

$$\sigma^* = \sigma/\cos\Theta = \sigma(1 + p_1^2 + p_2^2)^{1/2} \tag{11}$$

Both $\sigma$ and $\sigma^*$ are the functions of **n** (i.e. of $p_1$, $p_2$). The statistical properties of the surface of orientation $\mathbf{n}_0$ turn out to be closely related to the character of the $\sigma(\mathbf{n})$ – function in the vicinity of $\mathbf{n}_0$.

## 3.2 Nonsingular Surfaces

Let us consider the simplest case: the periodic terms $\psi(z_i)$in the general equation (9) are assumed to be constants. Then interface statistics is completely determined by interaction of the interface elements. If variation of $z_i$ along the interface is sufficiently slow, then it is possible to introduce a smooth profile function $z(x_1, x_2)$. The interaction free energy $\Psi_{int}$ depends on derivatives of z with respect to $x_\lambda$ ($\lambda = 1, 2$). The main contribution is due to the terms of the lowest order; they should appear in the form of scalar combination (with respect to the two-dimensional space of coordinates $x_1$, $x_2$)

$$A_\lambda \cdot \frac{\partial z}{\partial x_\lambda}; \qquad B_{\lambda\mu} \cdot \frac{\partial^2 z}{\partial x_\lambda \partial x_\mu}; \qquad \frac{1}{2}\sigma_{\lambda\mu} \cdot \frac{\partial z}{\partial x_\lambda} \cdot \frac{\partial z}{\partial x_\mu} \tag{12}$$

where $A_\lambda$ is a two-dimensional vector, whereas $B_{\lambda\mu}$ and $\sigma_{\lambda\mu}$ are two-dimensional tensors. Summation over repeated indices $\lambda$, $\mu$ is meant in (12) and further on. The terms (12) should be integrated over the area S. The first term yields a quantity that depends only on the positions of the peripheral elements and equals $A_\lambda p_\lambda S$ according to Eq. (10). The second term yields a non-essential edge effect. Hence $\Psi_{int}$ can be written in the form

$$\Psi_{int} = \int \frac{1}{2}\sigma_{\lambda\mu} \frac{\partial z}{\partial x_\lambda} \frac{\partial z}{\partial x_\mu} dx_1 dx_2 + A_\lambda p_\lambda S \tag{13}$$

The interface of orientation $\mathbf{n}_0$ (or, shortly, $\mathbf{n}_0$-interface) is assumed to be stable against deviations from the mean flat shape $\bar{z} = \text{const}$, that is, $\Psi_{int}$ should be increased by such deviations. It means that the principal values of the $\sigma_{\lambda\mu}$-tensor are positive i.e.

$$\sigma_{11} > 0; \qquad \sigma_{22} > 0; \qquad \sigma_{11}\sigma_{22} > \sigma_{12}^2 \tag{14}$$

It is natural to call $\sigma_{\lambda\mu}$ "the stability tensor" of $\mathbf{n}_0$-interface.

To come back to an arbitrary interface state (described by positions $z_i$) one should replace the integral in Eq. (13) by some quadratic form $Q(z_1, z_2, \ldots)$ which depends on the position differences between neighboring elements. We do not need to give an explicit expression for Q; the subsequent results will be based on a simple theorem: if a function $\Psi(z_1, z_2, \ldots)$ (which is a combination of quadratic form Q and any linear form) has an extremum at $z_i = z_i^0$ then

$$\Psi(z_1, z_2, \ldots) = \Psi(z_1^0, z_2^0, \ldots) + Q(\delta z_1, \delta z_2, \ldots) \tag{15}$$

where $\delta z_i = z_i - z_i^0$.

Particularly, if $z^0(x_1, x_2)$ is a slowly varying function then $\Psi_{int}(z_1^0, z_2^0, \ldots)$ can be found from Eq. (13). Besides, this expression yields the Euler's equation for $z^0(x_1, x_2)$

$$\sigma_{\lambda\mu} \frac{\partial^2 z^0}{\partial x_\lambda \partial x_\mu} = 0 \tag{16}$$

The distribution function of positions $z_i$ is given by Eq. (8). On substituting (15) into (8) one can see that $z^0(x_1, x_2)$ is identical to the mean interface position $\bar{z}(x_1, x_2)$.

### 3.2.1 *Angular Dependence of Specific Free Energy*

The minimum of $\Psi_{int}$ for **n**-interface is reached, according to Eq. (16), if the positions $z_i$ lie all in the same plane, that is, in the fixation plane of the peripheral elements. Substituting the general expression (15) into (8) and integrating over positions $z_i$ (except for the peripheral ones) one expresses the partition function of the **n**-interface through that of the $\mathbf{n}_0$-interface

$$Z_s(\mathbf{n}) = Z_s(\mathbf{n}_0) \exp\left[-\frac{1}{k_0 T} \Psi_{int}(z_1^0, z_2^0, \ldots)\right] \tag{17}$$

where $\Psi_{int}$ is determined by expr. (13) while $z^0$ is given by (10). Thus one finds the specific free energy $\sigma^* = -(k_0 T/S) \ln Z_s$

$$\sigma^*(\mathbf{n}) = \sigma(\mathbf{n}_0) + A_\lambda p_\lambda + \frac{1}{2} \sigma_{\lambda\mu} p_\lambda p_\mu \tag{18}$$

The quantity $\sigma^*$ (as well as $\sigma$) is a twice differentiable function of slopes $p_\lambda$ in the vicinity of $\mathbf{n}_0$. In this case the $\mathbf{n}_0$-interface is referred to as a "nonsingular surface" [6]. The stability tensor $\sigma_{\lambda\mu}$ can be expressed from (18) through the function $\sigma^*(p_1, p_2)$ or through $\sigma(p_1, p_2)$ if Eq. (11) is taken into account

$$\sigma_{\lambda\mu} = \frac{\partial^2 \sigma^*}{\partial p_\lambda \partial p_\mu} = \sigma \delta_{\lambda\mu} + \frac{\partial^2 \sigma}{\partial p_\lambda \partial p_\mu} \tag{19}$$

where all the quantities are taken at $p_\lambda = 0$. Slope $p_\lambda$ in (19) may be replaced by tilt angles $\Theta_\lambda = \arctan p_\lambda$, e.g.

$$\sigma_{11} = \sigma + \frac{\partial^2 \sigma}{\partial \Theta_1^2} \tag{20}$$

The stability conditions (14) impose some limitations on the $\sigma(\mathbf{n})$-function; the particular limitation $\sigma_{11} > 0$ was pointed out in Ref. 7.

Up to this point we dealt with surfaces, slightly tilted against the basic $\mathbf{n}_0$-surface. The role of basic surface can be played by any **n**-surface for which $\bar{z}_i$ varies according to Eq. (10). Then the expressions (12) should contain $z - \bar{z}$ instead of z while $\sigma_{\lambda\mu}$, $A_\lambda$, $B_{\lambda\mu}$ should be replaced by new quantities $\sigma^*_{\lambda\mu}$ etc. which characterize the **n**-surface. The "reduced stability tensor" $\sigma^*_{\lambda\mu}$ satisfies the inequalities (14). It is related to $\sigma^*$ by the formula similar to (19)

$$\sigma^*_{\lambda\mu}(p_1, p_2) = \frac{\partial^2 \sigma^*(p_1, p_2)}{\partial p_\lambda \partial p_\mu} \tag{21}$$

The tensor $\sigma^*_{\lambda\mu}$ coincides with $\sigma_{\lambda\mu}$ at $p_\lambda = 0$.

### 3.2.2 *Phase Equilibrium at Curved Interface*

If $f \neq 0$ then equilibrium distribution of interface positions is determined by the total partition function $Z_v \, dZ_s$. According to (4), (7), (8) multiplying $dZ_s$ by $Z_v$ is equivalent to replacing $\Psi$ by the effective quantity

$$\Psi_{ef} = \Psi - f q_c \Delta s \Sigma z_i \tag{22}$$

This combination of quadratic and linear forms reaches its minimum at some curved interface $z_i^0 = z^0(x_1, x_2)$ (the peripheral elements are supposed to be fixed along a smooth line). As the relation (15) is valid for $\Psi_{ef}$, the distribution of $z_i - z_i^0$ is the same as in case of flat mean shape at $f = 0$. Let us find the equation for a curved interface $\bar{z} = z^0(x_1, x_2)$ which is close to a plane defined by Eq. (10). The interaction free energy $\Psi_{int}$ is given by Eq. (13), if $\sigma_{\lambda\mu}$ is replaced by $\sigma^*_{\lambda\mu}$ and $\partial z/\partial x_\lambda$ is replaced by $(\partial z/\partial x_\lambda - p_\lambda)$. The second term in the right-hand part of (22) equals $-f q_c \int z \, dx_1 \, dx_2$ for a smooth shape $\bar{z}(x_1, x_2)$. Minimizing expr. (22) leads to the sought after equation for the equilibrium interface shape $\bar{z}(x_1, x_2)$

$$f q_c + \sigma^*_{\lambda\mu}(p_1, p_2) \frac{\partial^2 \bar{z}}{\partial x_\lambda \partial x_\mu} = 0 \tag{23}$$

where $p_\lambda = \partial \bar{z}/\partial x_\lambda$. Eq. (23) can be also derived[8] by minimizing the total free energy $\int (\sigma^* - f q_c) \, dx_1 \, dx_2$.

If a local coordinate system is introduced (the axis z being normal to interface at a given point while the axes $x_1$, $x_2$ coinciding with the principal axes of curved surface) then Eq. (23) takes the form

$$f q_c - \sigma_{11} \zeta_1 - \sigma_{22} \zeta_2 = 0 \tag{24}$$

where $\zeta_1$, $\zeta_2$ are the principal curvatures (assumed positive for a convex crystal surface) while the stability coefficients $\sigma_{11}$, $\sigma_{22}$ are defined by Eq. (20) and by the similar expression for $\sigma_{22}$. The equilibrium equation (24) was first derived by Herring[9]. The other form (23) of the equilibrium equation may appear to be more convenient due to the fixed coordinate system used for the whole interface.

### 3.2.3 Fluctuations of Interface

As fluctuations do not depend on the mean interface shape (flat or curved) we can put $f = 0$ and $\bar{z}_i = 0$. Let us find the mean square of position difference between the elements i and k, that is, the quantity $\overline{(z_i - z_k)^2}$. The general method to find the distribution function of a linear combination $u = \Sigma a_i z_i$ is the following. There exists a certain linear transformation from $z_i$ to some new variables $u_i$ (one of which equals u) that casts the quadratic form $Q(z_1, z_2, \ldots)$ into canonical form $\Sigma b_i u_i^2$. Thus Q depends on the variable u only due to a quadratic term $bu^2$. Consequently, the distribution function of u is proportional to $\exp(-bu^2/k_0 T)$. On the other hand, at a given u the free energy $\Psi = Q + \text{const}$ reaches some minimum value $\Psi_{min}(u)$ which is equal to $bu^2 + \text{const}$. Thus the distribution problem is reduced to calculating $\Psi_{min}(u)$.

The case we are interested in is $u = z_i - z_k$. If the functional (13) has a minimum at a given u then the extreme profile $z^0(x_1, x_2)$ satisfies Euler's equation (16) outside the elements i and k whereas $z^0 = \pm u/2$ at these elements. Let us direct the $x_\lambda$-axes along the principal axes of $\sigma_{\lambda\mu}$ and introduce new coordinates $x_1' = x_1/\sqrt{\sigma_1}$, $x_2' = x_2/\sqrt{\sigma_2}$ where $\sigma_1$, $\sigma_2$ are the principal values of $\sigma_{\lambda\mu}$. Then Eq. (16) takes the form of two-dimensional Laplace equation. Its solution is

$$z^0 = \frac{u}{2}\left(\ln\frac{r_k'}{r_i'}\right)\left(\ln\frac{r_{ik}'}{r_0'}\right)^{-1} \tag{25}$$

where $r_i'$ is the distance between a given surface point and the element i, $r_{ik}'$ is the distance between the two elements, $r_0'$ is the order of element size. Here a prime indicates that a quantity corresponds to the new coordinate system $x_1'$, $x_2'$. After coordinate transformation the integral (13) can be reduced to two linear integrals over circles enclosing the elements i and k. Taking into account Eq. (25) one obtains

$$\Psi_{min}(u) = \frac{\pi}{2} u^2 \sqrt{\sigma_1 \sigma_2}\left(\ln\frac{r_{ik}'}{r_0'}\right)^{-1} \tag{26}$$

where the additive constant is omitted. Since the distribution of u is given by $\exp(-\Psi_{min}/k_0 T)$, the mean square of u equals

$$\overline{u^2} = \overline{(z_i - z_k)^2} = \frac{k_0 T}{\pi\sqrt{\sigma_1 \sigma_2}} \ln\frac{r_{ik}'}{r_0'} \tag{27}$$

The distance $r_{ik}'$ can be written in the invariant form: $r_{ik}' = (\sigma_{\lambda\mu}^{-1}\Delta x_\lambda \Delta x_\mu)^{1/2}$ where $\Delta x_\lambda$ is the coordinate difference between the elements. The quantity $\overline{(z_i - z_k)^2}$ increases logarithmically with increasing the distance $r_{ik}$ between the two elements. It means that correlation length is infinite. The mean square of elements position, $\overline{z_i^2}$, is the order of (27) if $r'_{ik}$ is replaced by the distance $L'_i$ between the element and the edge of the interface. Thus $\overline{z_i^2}$ increases as the logarithm of the interface area S.

### 3.3 Singular Surfaces

In general case the interface free energy $\Psi$, given by Eq. (9), contains periodic terms $\psi(z_i)$. The function $\psi(z_i)$ reaches its minimum at some position $z_s$ and also at every equivalent position $z_s + jh$ where j is an integer. The periodicity of $\psi(z_i)$ does not yet mean that statistical properties of an interface are different from those considered above. Indeed, if shifting $z_i$ by h leads to small (in comparison to $k_0T$) change in $\Psi_{int}$ then periodic factors $\exp(-\psi/k_0T)$ can be averaged over the period h i.e. the free energy $\Psi(z_1, z_2, \ldots)$ is reduced to the previous form $\Psi_{int}$ + const. Let $\psi(z_i)$ have a deep sharp minimum, that is, $z_i$ is close to discrete values $z_s + jh$. The interaction free energy is again supposed to consist of quadratic terms $\varepsilon\,\Delta z^2$ where $\Delta z$ is the position difference between the neighboring elements. This case is identical to the discrete Gaussian model[10]. If the energy parameter $\varepsilon$ is less than some critical value $\varepsilon_{cr}$ then $\overline{(z_i - z_k)^2}$ increases logarithmically with increasing the distance $r_{ik}$ between the two elements – just as in the previous case of the constant $\psi$. Hence it is natural to conclude that the $\mathbf{n}_0$-surface is nonsingular at $\varepsilon < \varepsilon_{cr}$. On the other hand, if $\varepsilon > \varepsilon_{cr}$ then $\overline{(z_i - z_k)^2}$ tends to a certain finite limit at $r_{ik} \to \infty$, i.e. the correlation length is finite[10].

Let us now consider the general case of an interface characterized by a finite correlation length $l_c$; the mean square of position fluctuation, $\overline{(z_i - z_s)^2}$, is also finite. A portion of interface, consisting of n elements, has the area $s = n\,\Delta s$ and the position $z = (\Sigma z_i)/n$ (the sum is taken over n elements). If a linear size of this interface portion is greater than $l_c$ then the mean square of position fluctuation is inversely proportional to s

$$\overline{(z - z_s)^2} = B/s \tag{28}$$

where B is a constant. Therefore, if the interface is divided into sufficiently large elements of area s, then the element fluctuation is small in comparison with h. In other words all the elements are practically fixed at the same equilibrium position $z_s$. Of course there are equivalent states of the interface corresponding to the elements all having the same position $z_s + jh$. These discrete interface states can be formally treated as the different two-dimensional phases, the height $z_s + jh$ being an analog of phase density. The linear boundary between two such phases (of heights $z_s$ and $z_s + h$) has the sense of elementary surface step of height h. Step orientation is characterized by the unit normal **m** which is perpendicular to $\mathbf{n}_0$ and directed into the phase of lesser height. The tangential movement of a step (along **m**) corresponds to the growth of crystal monolayer. The transition from the initial one-phase surface state to the two-phase state with a step leads to some increase in the free surface energy $F_s$ which is proportional to the step length. The specific free energy of the step (per unit length) will be denoted by $\alpha(\mathbf{m})$.

Figure 3 shows a special interface configuration: terraces of heights $z_s + jh$ are separated by steps of the same orientation **m**. This stepped (vicinal) surface is tilted against the basic $\mathbf{n}_0$-surface by a small angle $\Theta = \arctan(h/Y)$ where Y is the average step spacing. On a large scale (greater than Y) the surface shape is described by equation

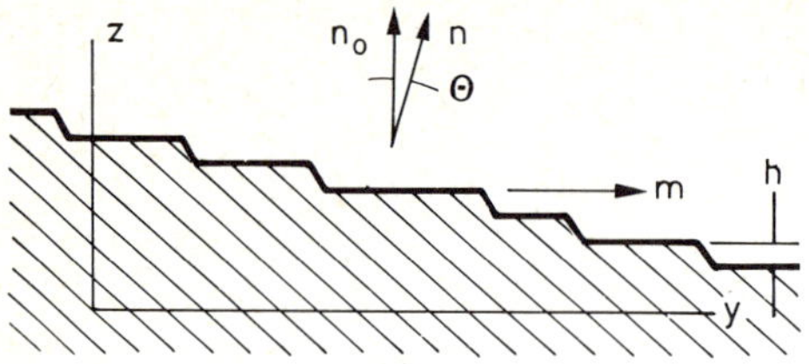


**Fig. 3.** A vicinal surface

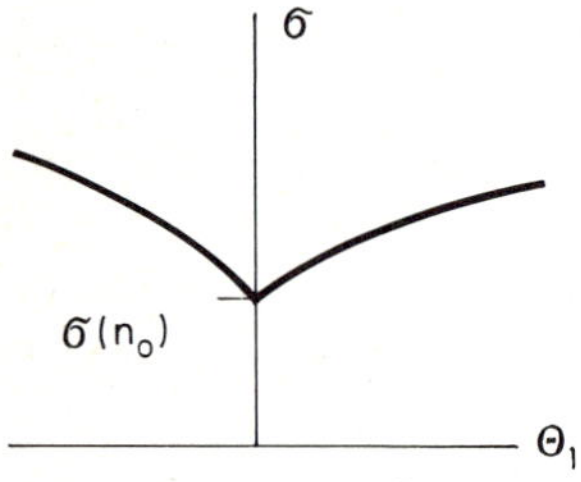


**Fig. 4.** Angular dependence of $\sigma$ in the vicinity of a singular orientation

$z = -m_\lambda x_\lambda h/Y$ + const where $m_\lambda$ is a component of **m** along the $x_\lambda$-axis ($\lambda = 1, 2$). Comparing this equation to Eq. (10) one finds the slopes $p_\lambda = -m_\lambda h/Y$. The free energy $F_s$ is equal to $\sigma(\mathbf{n}_0)\,S + \alpha(\mathbf{m})\,LN$ where L is the length of each step and N is the number of steps; the area S is equal to LNY. The average step spacing Y can be expressed through $p_\lambda$, then the specific free energy $\sigma^* = F_s/S$ equals

$$\sigma^*(\mathbf{n}) = \sigma(\mathbf{n}_0) + \frac{1}{h}\,\alpha(\mathbf{m})\sqrt{p_1^2 + p_2^2} \tag{29}$$

The function $\sigma^*(\mathbf{n})$ as well as $\sigma(\mathbf{n})$ exhibits singularity at $\mathbf{n} = \mathbf{n}_0$: an angular derivative (e.g. $\partial\sigma/\partial p_1$ at $p_2 = 0$) has a jump (Fig. 4) which is equal to $h^{-1}[\alpha(\mathbf{m}) + \alpha(-\mathbf{m})]$ where the unit normal **m** is directed along the $x_1$-axis. The $\mathbf{n}_0$-surface is called singular in this case[6)].

Different atomic models of an interface (discrete Gaussian model, Kossel model and its versions) lead to the conclusion[10–12)] that decreasing the energy of binding between surface atoms results in an increase of the correlation length $l_c$ and a decrease of the specific free energy $\alpha$; at some critical value of binding energy the two-dimensional phase transition takes place: $l_c$ becomes infinite, $\alpha$ becomes zero. This phase transition (usually called "roughening transition") is actually a singular-nonsingular transformation of the surface. It is evident that only a finite number of crystal surfaces is singular (this group includes only those surfaces which are parallel to the close-packed atomic planes characterized by sufficiently high energy of binding between neighboring atoms). In particular, all the surfaces may be nonsingular.

## 3.4 Layer Growth Mechanism

The elements of area s, that constitute a singular surface, are statistically independent (their size is greater than $l_c$). Hence the free energy $\Psi(z_1, z_2, \ldots)$ consists of quadratic contributions of all the elements; if the equilibrium position $z_s$ is zero then

$$\Psi = \frac{k_0 Ts}{2B} \Sigma z_i^2 + \text{const} \tag{30}$$

where the Eq. (28) is taken into account. The expansion (30) is valid until $|z_i|$ exceeds some limit value $z_{max}$. At non-zero driving force f the position distribution is given by $\exp(-\Psi_{ef}/k_0 T)$ where $\Psi_{ef}$ is defined by (22). The minimum of $\Psi_{ef}$ is reached at $z_i = \bar{z} = fq_c B/k_0 T$; if f is low enough then z remains less than $z_{max}$. According to the general relation (15) the distribution of deviations $z_i - \bar{z}$ is the same as at $f = 0$. So, at sufficiently low f the singular surface stays in a metastable state: all the elements have the same mean position $\bar{z}$, the fluctuation amplitude being rather small. Such metastable state exists up to some critical driving force $f_{cr}$; estimates of $f_{cr}$ were obtained for particular models of a singular interface[13–15].

At $f < f_{cr}$ the singular interface can advance (i.e. crystal growth can proceed) only by tangential movement of steps. Steps are generated either by two-dimensional nucleation (if the crystal is perfect) or at dislocations and other defects. The generation rate, sufficient for the crystal to grow at a given normal rate V is achieved only at appreciable f. It is therefore likely that the tangential step velocity v is much higher than V, which means that the average step spacing $Y = vh/V$ is much greater than the step height h. Then a surface tilted against the singular $\mathbf{n}_0$-surface by a small but appreciable angle (say $\gtrsim 1°$) has a comparatively high step density and so requires extremely low driving force f to advance at the same rate V. As a result the total interface consists of regions of two types:
A) a curved nonsingular interface that practically coincides with the solidification isotherm $T_0$ or, more strictly, satisfies the equilibrium Eq. (24);
B) almost flat nonisothermal facets which are practically parallel to singular surfaces.

For instance, Fig. 5a shows a shape of crystallization front, typical for Czochralski pulling process (the case of convex solidification isotherm). Step generation is usually located in the vicinity of the facet's coldest point $\mathcal{M}$. Outside this generation zone the facet is actually a vicinal surface of extremely low tilt angle $\Theta = V/v$. A nonsingular interface of appreciable tilt (e.g. the part $\mathcal{EL}$ in Fig. 5a) coincides with the isotherm $T_0$ except for a small vicinity of the three-phase line $\mathcal{L}$ where the interface is strongly curved[16, 17] and so deviates from the isotherm according to Eq. (24). The existence of singular orientations tells essentially not only on the shape of crystallization front but also on the external shape of pulled crystals[18, 19].

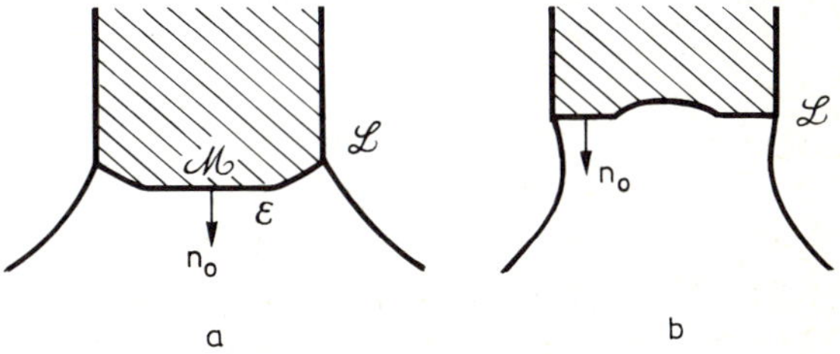


**Fig. 5a, b.** Faceting of the crystallization front of a pulled crystal

## 4 Thermodynamics and Kinetics of Steps

### 4.1 Description of Step States

A step can be described by the approach quite similar to that used for a surface. An atomic plane $\mathcal{P}$, parallel to the singular surface under consideration, consists of parallel atomic rows $\mathcal{R}_j$; the unit normal to these rows is denoted by $\mathbf{m}_0$. A step of unit normal $\mathbf{m}$ (which is close to $\mathbf{m}_0$) separates the two-dimensional phases of heights $z_s$ and $z_s + h$. Let us distinguish a small rectangular portion of the surface that contains a portion of the step; the rectangle base is fixed with respect to one of the $\mathcal{R}_j$-rows. The geometry is defined by Fig. 1 if one replaces $\mathcal{P}_j$ by $\mathcal{R}_j$ and the z-axis by the y-axis directed along $\mathbf{m}_0$ (the x-axis is directed along $\mathcal{R}_j$). The rectangle sizes along the x- and y-axis are $\Delta l$ and $w_e$, respectively; the size $w_e$ should exceed the correlation length $l_c$ so that the surface element can be treated as composed of two phases (of heights $z_s$ and $z_s + h$). The position z of the rectangular surface elements is intermediate between $z_s$ and $z_s + h$ and can serve as a measure of step advance along $\mathbf{m}_0$. The position y of a given step element can be defined by the relation similar to (1)

$$(z - z_s)\, w_e = (y - y_b)\, h \tag{31}$$

where $y_b$ is the coordinate of the rectangle base.

The quantity z changes not only due to step motion but also to fluctuations of the position of each of two-dimensional phases; the fluctuation amplitude equals $(B/w_e\,\Delta l)^{1/2}$ according to (28). Hence the step element position y has an uncertainty

$$\delta y \sim \frac{1}{h}\left(\frac{Bw_e}{\Delta l}\right)^{1/2} \tag{32}$$

At a sufficiently large ratio $\Delta l/w_e$ the uncertainty is much smaller than the row spacing $h_e$, that is, the quantity y is suitable for describing infinitesimal displacements of the step element.

A step that consists of $\nu + 1$ elements is described by the set of element positions $y_i$ $(i = 0, 1, \ldots, \nu)$. At $f = 0$ the partition function of a surface with a step (the positions $y_i$ are confined to intervals $dy_i$) differs from that of a surface without step by a factor

$$dZ_e = \exp\left[-\frac{1}{k_0 T}\,\Psi(y_0, \ldots, y_\nu)\right]\frac{dy_1}{a}\ldots\frac{dy_{\nu-1}}{a} \tag{33}$$

Now $\Psi$ has the sense of the step free energy for a given set of positions $y_i$. It includes periodic terms $\psi(y_i)$ of the period $h_e$ and the interaction term $\Psi_{int}$ which depends on the position differences. The positions $y_0$ and $y_\nu$ of the two extreme elements are fixed so that a step orientation $\mathbf{m}$ is defined.

If $\psi(y_i)$ has a deep sharp minimum at $y_i = y_e$ (as well as at the equivalent positions $y_e + jh_e$) then a step consists of straight segments separated by kinks. The linear density of kinks, $\varrho$, is finite at any temperature $T > 0$[20, 21]. The position difference $\Delta y$ between two elements, separated by a distance $\Delta x$, equals $h_e(N_+ - N_-)$ where $N_+$ and $N_-$ are the numbers of "positive" and "negative" kinks, respectively, situated between the two elements. Since $N_+$ and $N_-$ obey Poisson's distribution law, the amplitude of fluctuations of $\Delta y$ equals $h_e(\varrho \Delta x)^{1/2}$ and becomes large in comparison with $h_e$ if a segment $\Delta x$ contains many kinks. Thus if a step is divided into sufficiently large elements (of length $\Delta l \gg \varrho^{-1}$) then the relative displacement of two adjacent fluctuating elements is much greater than $h_e$. Therefore the element properties can be averaged over the period $h_e$; after that $\psi(y_i)$ is replaced by some constant $\psi$, i.e. the dependence of $\Psi$ on positions $y_i$ is determined only by $\Psi_{int}$. In other words, a step (unlike a surface) is always nonsingular. The absence of singular steps is related to impossibility of different phases of infinite length in a one-dimensional system[2].

## 4.2 Specific Free Energy

The step free energy $F_e$ is equal to $-k_0 T \ln Z_e$ where the partition function $Z_e$ is obtained by integrating Eq. (33) over the positions $y_1, \ldots, y_{\nu-1}$. Step orientation may be specified by the unit normal **m**, by the tilt angle $\varphi$ (i.e. the angle between **m** and $\mathbf{m}_0$) or by the slope $p = \tan\varphi = (y_\nu - y_0)/L$ where $L = \nu \Delta l$ is a step length along the x-axis (the total length $L_t$ equals $L/\cos\varphi$). Beside the specific free energy $\alpha = F_e/L_t$ it is convenient to introduce the reduced quantity $\alpha^* = F_e/L$

$$\alpha^* = \alpha/\cos\varphi = \alpha\sqrt{1 + p^2} \tag{34}$$

If variation of $y_i$ along the step is slow enough then it is possible to introduce a smooth profile function $y(x)$. The interaction free energy $\Psi_{int}$ depends on derivatives of $y(x)$, the main role being played by the terms of lowest order

$$A_e \frac{dy}{dx}; \qquad B_e \frac{d^2y}{dx^2}; \qquad \frac{1}{2}\alpha_e\left(\frac{dy}{dx}\right)^2 \tag{35}$$

Integration of these terms over x yields the expression similar to (13)

$$\Psi_{int} = \int_0^L \frac{1}{2}\alpha_e\left(\frac{dy}{dx}\right)^2 dx + A_e p L \tag{36}$$

The quantity $\alpha_e > 0$ will be called "stability coefficient" of the $\mathbf{m}_0$-step. For a general step state described by positions $y_i$ the integral in Eq. (36) should be replaced by quadratic form. Then

$$\Psi = \Psi_{int} + \nu\psi = \frac{\alpha_e}{2\Delta l}\sum_{i=1}^{\nu}(y_i - y_{i-1})^2 + \nu(\psi + A_e p \Delta l) \tag{37}$$

Since a step is a one-dimensional analog of a nonsingular surface, the results obtained in subsections 3.2.1 and 3.2.2 can be almost directly applied to the case of a step. The angular dependence of $\alpha^*$ at a small slope is determined by an expression similar to (18)

$$\alpha^*(\mathbf{m}) = \alpha(\mathbf{m}_0) + A_e p + \frac{1}{2}\alpha_e p^2 \tag{38}$$

Thus the stability coefficient $\alpha_e$ can be expressed through either $\alpha^*(p)$ or $\alpha(\varphi)$

$$\alpha_e(\mathbf{m}_0) = \frac{d^2\alpha^*}{dp^2} = \alpha + \frac{d^2\alpha}{d\varphi^2} \tag{39}$$

where all the quantities are taken at $p = 0$.

Small deviations of element positions $y_i$ from an arbitrary tilted straight profile $\bar{y}_i = px + y_0$ are characterized by the reduced stability coefficient $\alpha_e^*$ which is related to $\alpha^*(p)$ by the expression similar to (21)

$$\alpha_e^*(p) = \frac{d^2\alpha^*(p)}{dp^2} \tag{40}$$

The relation between $\alpha_e(\mathbf{m})$ and $\alpha_e^*(p)$ for a given tilted step can be derived as follows. The second of the two expressions (39) is valid not only for the $\mathbf{m}_0$-step but also for an arbitrary **m**-step since this expression is invariant under rotation of the coordinate system x, y. Expressing $\alpha(\varphi)$ through $\alpha^*(p)$ from (34) and performing the differentiation one obtains

$$\alpha_e(\mathbf{m}) = (1 + p^2)^{3/2}\alpha_e^*(p) \tag{41}$$

At $f \neq 0$ the mean step profile $\bar{y}(x)$ is curved; the equilibrium distribution of positions is given by Eq. (33) if $\Psi$ is replaced by the effective quantity $\Psi_{ef}$ according to Eq. (22). A step contribution to $\Delta s \Sigma z_i$ (crystal volume) is equal to the area of a layer behind the step, $\Delta l \Sigma y_i$, multiplied by the step height h. Hence

$$\Psi_{ef} = \Psi - fq\Delta l \Sigma y_i \tag{42}$$

where $q = q_c h$ is the area density of atoms in a crystal monolayer. The mean curved profile $\bar{y}(x)$ corresponds to the minimum of $\Psi_{ef}$ and satisfies the equation which is similar to (23)

$$qf + \alpha_e^*(p)\frac{d^2\bar{y}}{dx^2} = 0 \tag{43}$$

where $p = d\bar{y}/dx$. Taking into account Eq. (41) one can rewrite the equilibrium Eq. (43) in the invariant form

$$qf - \alpha_e\zeta = 0 \tag{44}$$

which is similar to (24). Here $\zeta$ is a step curvature assumed positive for a convex profile.

To describe an arbitrary **m**-step one can use a local coordinate system with the y-axis directed along **m**. The parameters $\Delta l$, $\psi$, $\alpha_e$ which correspond to this system are the functions of **m**. To express $\alpha(\mathbf{m})$ through these initial parameters one should find the partition function $Z_e$ by integrating (33) over $y_1, \ldots, y_{\nu-1}$ at $y_0 = y_\nu = 0$; the free energy $\Psi$ is given by Eq. (37) at $p = 0$. Evidently $Z_e$ is unchanged if multiplied by delta-function $\delta(y_\nu) = (2\pi)^{-1} \int \exp(j\varkappa y_\nu)\, d\varkappa$ and then integrated over $y_\nu$ (here j is imaginary unit). If new variables $u_i = y_i - y_{i-1}$ ($i = 1, \ldots, \nu$) are used then $Z_e$ is transformed to the product of $\nu$ identical Gaussian integrals which is to be integrated over $\varkappa$. Performing the integration and omitting an inessential constant factor one obtains

$$Z_e = \left(\frac{2\pi k_0 T \Delta l}{\alpha_e a^2}\right)^{\nu/2} \left(\frac{k_0 T}{\alpha_e L}\right)^{1/2} \exp\left(-\frac{\nu\psi}{k_0 T}\right) \tag{45}$$

The step free energy $F_e = -k_0 T \ln Z_e$ consists of the principal term, proportional to $\nu$ (i.e. to a step length $L = \nu \Delta l$), and the correction term $0.5\, k_0 T \ln(\alpha_e L/k_0 T)$. The latter is much smaller than the former at sufficiently large L. The sought for expression for $\alpha = F_e/L$ is

$$\alpha = \frac{1}{\Delta l}\left(\psi - \frac{k_0 T}{2} \ln \frac{2\pi k_0 T \Delta l}{\alpha_e a^2}\right) \tag{46}$$

The parameters $\alpha$ and $\alpha_e$ are step characteristics that do not depend on division scale $\Delta l$. Therefore the Eq. (46) is essentially a relation between two division parameters $\Delta l$ and $\psi$.

## 4.3 Equilibrium Fluctuations

To find the distribution of position $y_i$ of a given step element one should calculate the minimum value of $\Psi(y_0, \ldots, y_\nu)$ at a given $y_i$ (according to the general method described in subsection 3.2.3). The profile $y_k^0$ that corresponds to the minimum of $\Psi$ satisfies the equation $\partial\Psi/\partial y_k = 0$ at $k \neq i$. Substituting $\Psi$ from (37) one gets $y_k^0 - y_{k-1}^0 = y_{k+1}^0 - y_k^0$, that is, the profile is straight outside the element i (Fig. 6). Then either Eq. (36) or expr. (37) yields

$$\Psi_{min}(y_i) = \frac{\alpha_e L}{2 x_i (L - x_i)} y_i^2 + \text{const} \tag{47}$$

Since the distribution of $y_i$ is given by $\exp(-\Psi_{min}/k_0 T)$ then

$$\overline{y_i^2} = \frac{k_0 T x_i (L - x_i)}{\alpha_e L} \tag{48}$$

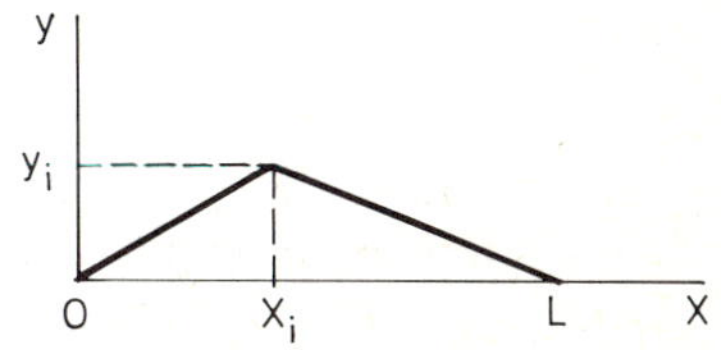


**Fig. 6.** The extremal step profile at a given position of the i-th step element

The fluctuation amplitude is independent of scale $\Delta l$ and proportional to $\sqrt{L}$ for the middle of a step. Quite similarly, $\Psi_{min}(y_i, y_k)$ can be calculated; this quantity determines the binary distribution. The extreme profile is now straight outside the elements i and k, hence

$$\Psi_{min}(y_i, y_k) = \frac{\alpha_e}{2}\left[\frac{y_i^2}{x_i} + \frac{(y_k - y_i)^2}{x_k - x_i} + \frac{y_k^2}{L - x_k}\right] \tag{49}$$

at $i < k$. Minimizing (49) at a given $y_i - y_k$ one can obtain the mean square of position difference between two elements

$$W_{ik} \equiv \overline{(y_i - y_k)^2} = \frac{k_0 T}{\alpha_e} |x_k - x_i| \tag{50}$$

if the distance $|x_k - x_i|$ is much smaller than L. As $W_{ik}$ increases with increasing $|x_k - x_i|$, the correlation length for a step is infinite.

## 4.4 Mean Velocity

If a step is characterized by certain positions $y_i$ at a moment t, then positions $y_i'$ at a moment $t + dt$ will be random variables. The mean displacement of a given element i is proportional to dt

$$\langle y_i' - y_i \rangle = v_i \, dt \tag{51}$$

Averaging with respect to new state at fixed initial state is denoted by triangle brackets (complete averaging at a given moment is denoted by a line, e.g. $\overline{y_i'}$). The quantity $v_i$ is the average velocity of the i-th element for a given state of the step. Beside the average displacements (51), independent random displacements of the elements take place. The mean square of element displacement is equal to $2D\,dt$ where D is the diffusion coefficient of a step element. Thus

$$\langle (y_i' - y_i)(y_k' - y_k) \rangle = 2D\,\delta_{ik}\,dt \tag{52}$$

The probability of a given state (characterized by positions confined between $y_i$ and $y_i + dy_i$) is equal to $C(y_1, \ldots, y_{\nu-1}) \cdot dy_1 \ldots dy_{\nu-1}$. The probability density C can be treated as the concentration of points in the multidimensional space of variables $y_i$. The flux $J_i$ of these points along the $y_i$-axis is determined by their drift and diffusion

$$J_i = v_i C - D \frac{\partial C}{\partial y_i} \tag{53}$$

The equilibrium distribution $C \sim \exp(-\Psi_{ef}/k_0 T)$ corresponds to zero flux. Then Eq. (53) yields the Einstein's relation between the velocity $v_i$ and the diffusivity D

$$v_i = -\frac{D}{k_0 T}\frac{\partial \Psi_{ef}}{\partial y_i} \tag{54}$$

Making use of expressions (37), (42) one obtains

$$v_i = \frac{D}{k_0 T}\left[fq\,\Delta l + \frac{\alpha_e}{\Delta l}(y_{i+1} + y_{i-1} - 2y_i)\right] \tag{55}$$

This equation should be averaged with respect to initial state of the step. According to (51) $\bar{v}_i = \partial\bar{y}_i/\partial t$. If the mean step profile is straight (i.e. all $\bar{y}_i$ are the same) then the mean normal velocity of the step (denoted by v) is proportional to the driving force f, that is, to the supercooling $\Delta T$

$$v = \beta\frac{f}{k_0 T} = \beta_T \Delta T \tag{56}$$

where $\beta = Dq\,\Delta l$. The relation between the kinetic coefficients of a step, $\beta$ and $\beta_T$, follows from (6): $\beta_T = \beta H/k_0 T^2$. Since $\beta$ is a step property that does not depend on a scale $\Delta l$, it is convenient to express D through $\beta$ and $\Delta l$

$$D = \frac{\beta}{q\,\Delta l} \tag{57}$$

If the mean step profile $\bar{y}_i$ is curved, but $\bar{y}_i$ is a slow function of i which can be replaced by a smooth profile $\bar{y}(x, t)$ then the averaged Eq. (55) takes the form

$$v = \frac{\beta}{k_0 T}\left(f - \frac{\alpha_e}{q}\zeta\right) \tag{58}$$

where $\zeta = -d^2\bar{y}/dx^2$ is the curvature of mean profile. Due to invariant form of Eq. (58) one can use it for a step tilted against the x-axis by an arbitrary angle $\varphi$. For such a step $\partial\bar{y}/\partial t = v/\cos\varphi$; if the curvature $\zeta$ is expressed through $\bar{y}(x, t)$ and relation (41) is used, then kinetic Eq. (58) takes the form that explicitly determines the evolution of the mean step profile

$$\frac{\partial\bar{y}}{\partial t} = \frac{\beta^*}{k_0 T}\left(f + \frac{\alpha_e^*}{q}\frac{\partial^2\bar{y}}{\partial x^2}\right) \tag{59}$$

where $\beta^* = \beta\sqrt{1 + p^2}$ is the reduced kinetic coefficient and $p = \partial\bar{y}/\partial x$. Equation (59) has a form of diffusion equation; if a slope p is small enough then the "diffusion coefficient" is constant and equal to

$$D_e = \frac{\beta\alpha_e}{k_0 Tq} \tag{60}$$

Step movement is accompanied by heat production, equal to qH per unit area of crystal monolayer. Therefore the supercooling $\Delta T$, which has been assumed to be an independent parameter, is actually a function of v. However, if the heat conductivity of either the crystal or the melt is much greater than $qH\beta_T$, the effect of heat production on $\Delta T$ is negligible[22].

The kinetic Eqs. (58), (59) could be written on the basis of the linear growth law (56) for a straight step and the equilibrium Eq. (43) or (44) for a curved step. The similar considerations lead to the kinetic equation for nonsingular surface if the equilibrium Eq. (23) is taken into account

$$\frac{\partial \bar{z}}{\partial t} = \frac{\mathcal{B}^*}{k_0 T}\left(f + \frac{1}{q_c}\sigma^*_{\lambda\mu}\frac{\partial^2 \bar{z}}{\partial x_\lambda \partial x_\mu}\right) \tag{61}$$

where $\mathcal{B}^*$ is the reduced kinetic coefficient of a surface ($\mathcal{B}^* = \mathcal{B}/\cos\Theta$ where $\Theta$ is a tilt angle and $\mathcal{B}$ is the surface kinetic coefficient, that is, the proportionality coefficient between the normal growth rate and $f/k_0 T$ for a flat interface). Kinetic equations, similar to (59), (61), could be also written on the basis of the general principle of proportionality between the growth rate and the variational derivative of free energy F with respect to displacement of a step or a nonsingular surface[23].

## 4.5 Fluctuations of a Moving Step

The evolution of the mean step profile is completely determined by the averaged Eq. (55). Let us now examine deviations of $y_i$ from $\bar{y}_i(t)$. Performing multiplication in the lefthand part of Eq. (52) and taking into account Eq. (51) we get

$$\langle y_i' y_k' - y_i y_k\rangle = (2D\delta_{ik} + y_i v_k + y_k v_i)\,dt \tag{62}$$

If this equation is averaged with respect to initial state of step and divided by dt, then the left-hand part is equal to $d\,(\overline{y_i y_k})/dt$. Each of the quantities $y_i$, $y_k$, $v_i$, $v_k$ in the averaged Eq. (62) can be expressed through the mean value and the deviation ($y_i = \bar{y}_i + \delta y_i$ etc.). Making use of Eq. (55) and the relation $\partial\bar{y}_i/\partial t = \bar{v}_i$ we obtain the equation for the correlation function $Q_{ik} = \overline{\delta y_i \delta y_k}$

$$\frac{dQ_{ik}}{dt} = 2D\delta_{ik} + \frac{D\alpha_e}{k_0 T \Delta l}(Q_{i+1k} + Q_{i-1k} + Q_{ik+1} + Q_{ik-1} - 4Q_{ik}) \tag{63}$$

If at $t = 0$ all the positions have certain values $y_i(0)$ then $Q_{ik}(0) = 0$. Due to homogeneity of this initial condition the quantity $Q_{ik}(t)$ is a function of the index difference, $i - k$. Then, putting $k = 0$ in Eq. (63) and denoting $Q_{i0}$ by $Q_i$, we rewrite Eq. (63) in the form

$$\frac{dQ_i}{dt} = 2D\delta_{i0} + \frac{2D\alpha_e}{k_0 T \Delta l}(Q_{i+1} + Q_{i-1} - 2Q_i) \tag{64}$$

The solution of Eq. (64) can be obtained by means of recursive relation for Bessel functions $I_i$ of imaginary argument[24]

$$Q_i = \frac{k_0 T \Delta l}{2\alpha_e} \int_0^{\vartheta} \exp(-\vartheta') I_i(\vartheta') \, d\vartheta' \tag{65}$$

where $\vartheta = 4 D \alpha_e t / k_0 T \Delta l$. At $\vartheta \gg 1$ one can use the asymptotic expression $I_i(\vartheta) = (2\pi\vartheta)^{-1/2} \exp(\vartheta - i^2/2\vartheta)$ which is valid at $|i| \ll \vartheta$. Substituting D from (57) and resuming the previous notation $Q_{ik}$ we obtain

$$Q_{ik} = \left(\frac{2\beta k_0 T t}{\pi q \alpha_e}\right)^{1/2} [\exp(-X^2) - \pi^{1/2} X \operatorname{erfc}(X)] \tag{66}$$

where $X = |x_i - x_k| / \sqrt{8 D_e t}$. The solution (66) could be found immediately from Eq. (64) if $Q_i$ were considered a smooth function of $x = i \Delta l$. In that case Eq. (64) is reduced to diffusion equation with a constant source at $x = 0$, the diffusivity being equal to $2 D_e$. The solution of this diffusion problem[25] coincides with (66). Correlation of positions $y_i$ and $y_k$ vanishes at distance $|x_i - x_k| \gtrsim \sqrt{8 D_e t}$. Thus the diffusion coefficient $D_e$ defined by (60) determines both evolution of the mean step profile and propagation of correlation along the step. The mean square of relative displacement $W_{ik} = \overline{(\delta y_i - \delta y_k)^2}$ equals $2(Q_{00} - Q_{ik})$. According to Eq. (66), $W_{ik}$ coincides with the equilibrium value (50) if $|x_i - x_k|$ is less then the propagation length of correlation $\sqrt{8 D_e t}$.

The important characteristic of a step is the fluctuation amplitude $y_f = (\overline{\delta y_i^2})^{1/2} = \sqrt{Q_{00}}$. At small t it follows from (65) that $y_f = \sqrt{2 D t}$ in accordance with the basic relation (52). But if correlation is spread over many elements of the step (i.e. $\sqrt{8 D_e t} \gg \Delta l$ and hence $\vartheta \gg 1$) then it follows from (66) that

$$y_f = (\chi t)^{1/4} \tag{67}$$

where

$$\chi = \frac{2\beta k_0 T}{\pi q \alpha_e} \tag{68}$$

Thus the random displacement of a step, that is, of a system of interacting elements, is proportional to $t^{1/4}$.

The total displacement of a given step element after a time interval t consists of the average displacement vt and the random one $\sim y_f$. The step will definitely advance from the initial position only at $t > (\chi/v^4)^{1/3}$ when $y_f$ is less than vt. Thus there exists a characteristic fluctuation time $\tau_f$; within that time a step fluctuates near the initial position. According to (56), (68)

$$\tau_f \sim \beta^{-1} \left(\frac{\alpha_e q}{k_0 T}\right)^{-1/3} \left(\frac{f}{k_0 T}\right)^{-4/3} \tag{69}$$

At $t \lesssim \tau_f$ a step can retreat from the initial position by the distance of order $y_f - vt$; the largest retreat displacement $y_r$ is reached at $t \sim \tau_f/6$ and is equal to

$$y_r \sim \frac{1}{2}\left(\frac{\alpha_e q f}{k_0^2 T^2}\right)^{-1/3} \tag{70}$$

The Eqs. (69), (70) were earlier obtained[26] from simple qualitative considerations.

## 4.6 Steps with Low Density of Kinks

If the $\mathbf{m}_0$-step (that is, the step which is parallel to the close-packed atomic rows $\mathcal{R}_j$) has a low equilibrium density of kinks, $\varrho_0$, then the interaction of kinks can be neglected and only kinks of the minimum depth $h_e$ may be taken into account. Then a step, tilted against the $\mathbf{m}_0$-step by a small angle $\varphi = \arctan p$ has the kink density $\varrho(p)$ and the specific free energy $\alpha(p) \approx \alpha^*(p)$ given by the equations[27]

$$\varrho(p) = [\varrho_0^2 + (p/h_e)^2]^{1/2} \tag{71}$$

$$\alpha(p) = \alpha(0) + k_0 T\left[\varrho_0 - \varrho + \frac{p}{h_e} \ln\left(\frac{\varrho}{\varrho_0} + \frac{p}{h_e \varrho_0}\right)\right] \tag{72}$$

Substituting Eq. (72) into (40) one obtains the stability coefficient of a step with low $\varrho$

$$\alpha_e \approx \alpha_e^* = \frac{k_0 T}{\varrho h_e^2} \tag{73}$$

The kinetic coefficient is proportional to the kink density[28]

$$\beta \approx \beta^* = d_e \varrho q h_e^2 \tag{74}$$

where $d_e$ is the diffusion coefficient of a kink.

Equations (71), (73), (74) lead to significant simplification of the kinetic Eq. (59) which can be then solved and used to analyze a spiral source of steps[27].

The quantities $D_e$, $\chi$, $\tau_f$, $y_r$ that characterize step fluctuations, can be now expressed through the kink parameters $\varrho$, $d_e$. Particularly, the diffusion coefficient $D_e$ which characterizes correlation propagation along a step, coincides with the kink diffusivity $d_e$. The retreat displacement (70) takes the form

$$y_r \sim h_e (\varrho x_r)^{1/3} \tag{75}$$

where $x_r = k_0 T/f q h_e$ is the retreat displacement of a kink[28].

If $x_r$ is much greater than the average kink spacing $1/\varrho$ then $y_r$ is large in comparison with $h_e$. It enables us to treat the process of capture of impurity atoms by a moving step by means of the collective parameters $\beta$, $\alpha_e$ instead of detailed analysis of kink motion[26, 29].

## 4.7 Vicinal Surfaces

A vicinal surface (Fig. 3) may be thermodynamically unstable i.e. elementary steps of the height h may unite, forming steps of greater height or even macrosteps[30, 31]. We shall confine ourselves to the simplest case of elementary steps; the step interaction is reduced to a purely geometrical limitation: each step is located between the two adjacent steps. The step train is usually analyzed[21, 32–34] under the assumption that each step is straight and moves at a certain velocity depending on the distances to the adjacent steps. Actually steps fluctuate, so any initially ordered structure of a vicinal surface will become chaotic (Fig. 7) when the random displacement (67) becomes the order of the average step spacing $Y = h/\Theta$ where $\Theta$ is a small tilt angle. Thus the randomization time is

$$\tau_d \sim Y^4/\chi = \left(\frac{\alpha_e q}{\beta k_0 T}\right)\left(\frac{h}{\Theta}\right)^4 \tag{76}$$

A step drifts over the characteristic distance Y within a certain time $\tau_v$ which is also the time for the interface to advance over the distance h, i.e.

$$\tau_v \sim h/V \tag{77}$$

where V is the normal growth rate.

The character of step motion depends on the ratio $\tau_v/\tau_d$.

If $\tau_v \ll \tau_d$ then the random displacement of a step within the time $\tau_v$ is relatively small; this condition can be written in a more precise form $(\chi \tau_v)^{1/4} \ll Y$ or $(\tau_v/\tau_d)^{1/4} \ll 1$. In this case each step has a practically definite velocity $v = \beta f/k_0 T$ which is independent of the adjacent steps. The normal growth rate V is equal to $v\Theta$, hence the kinetic coefficient of a vicinal surface $\mathcal{B} \approx \mathcal{B}^*$ is equal to $\beta \Theta$.

If $\tau_v \gg \tau_d$ then each step performs many random oscillations from one adjacent step to the other within the time $\tau_v$. In this case the motion of steps is a collective process.

To distinguish between the two regimes of step motion it is convenient to define the characteristic tilt angle $\Theta_0$ by the equation $\tau_v = \tau_d$. According to (76) and (77)

$$\Theta_0 = \left(\frac{\alpha_e q h^3 V}{\beta k_0 T}\right)^{1/4} \tag{78}$$

Substituting the silicon step parameters[22] and the typical growth rate $V \sim 3 \times 10^{-3}$ cm/s we obtain $\Theta_0 \approx 0.02$ (i.e. about 1°). Since the dependence of $\Theta_0$ on the parameters involved in (78) is rather weak, the estimate $\Theta_0 \sim 1°$ must be typical for

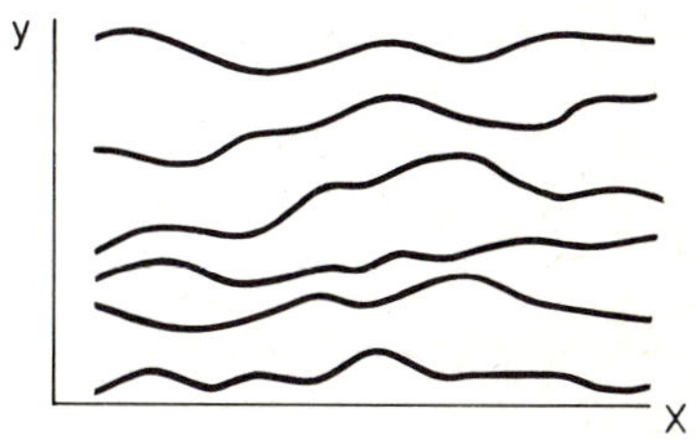


**Fig. 7.** Step distribution on a vicinal surface after randomization time $\tau_d$

crystal-melt interface. The regime of "directed step motion" ($\Theta \ll \Theta_0$) is realized for facets while the regime of "collective fluctuations" ($\Theta^4 \gg \Theta_0^4$) is realized for off-facet interface regions (e.g. the region $\mathscr{E}\mathscr{L}$ in Fig. 5 a). The growth rate $V = \partial \bar{z}/\partial t$ for the latter case is determined by the general Eq. (61). It is therefore necessary to calculate the tensor $\sigma^*_{\lambda\mu}$ for a vicinal surface.

To solve this problem one should take into account, that the specific free energy of a step, located on a vicinal surface, differs from $\alpha$ by some correction $\delta\alpha$ due to limitation on possible positions of a given step. To estimate $\delta\alpha$ we shall assume that a step fluctuates inside a region of width 2 Y i.e. between the two straight adjacent steps. Let us calculate the partition function $Z_\nu(y)$ of a step consisting of $\nu$ elements, the extreme right element having a certain position y. The quantity $Z_\nu(y)$ is obtained by integrating Eq. (33) over $y_1, \ldots, y_{\nu-1}$ between the limits $-Y$ and Y. Substituting Eq. (37) into (33) one gets a recursive relation

$$Z_{\nu+1}(y) = \int_{-Y}^{Y} Z_\nu(y') \; \exp\left[-\frac{\psi}{k_0 T} - \frac{\alpha_e}{2 k_0 T \Delta l}(y'-y)^2\right] a^{-1} dy' \tag{79}$$

at $A_e = 0$ (if $A_e \neq 0$ then $Z_\nu(y)$ is multiplied by an inessential factor $\exp(-A_e y/k_0 T)$). If the notation $t_e = k_0 T \Delta l / 2\alpha_e$ is introduced and $\psi$ is expressed from (46), then the relation (79) takes the following form

$$Z_{\nu+1}(y) = e^{-\frac{\alpha \Delta l}{k_0 T}} \int_{-Y}^{Y} Z_\nu(y')(4\pi t_e)^{-1/2} e^{-\frac{(y'-y)^2}{4 t_e}} dy' \tag{80}$$

The integral in the right-hand part of (80) can be considered to be the solution of diffusion equation $\partial c/\partial t = \partial^2 c/\partial y^2$ for an infinite straight line, the initial condition being $c = Z_\nu(y)$ at $|y| \leq Y$ and $c = 0$ at $|y| > Y$. If Y is much larger than the element size $\Delta l$, then the "diffusion length" $\sqrt{2 t_e}$ is small in comparison with Y. Then the change of the "concentration" c within the "time interval" $t_e$ is small, except for the narrow regions of width $\sim \sqrt{2 t_e}$ near $\pm Y$. Thus $c \approx Z_\nu + t_e \partial c/\partial t$ at $t = t_e$. Expressing $\partial c/\partial t$ (at $t = 0$) from the diffusion equation one obtains

$$Z_{\nu+1} = e^{-\frac{\alpha \Delta l}{k_0 T}} \left(Z_\nu + t_e \frac{d^2 Z_\nu}{dy^2}\right) \tag{81}$$

At $\nu \gg 1$ the stationary distribution is established, that is, $Z_{\nu+1}(y)$ becomes proportional to $Z_\nu(y)$. Then it follows from (81) that $\partial^2 Z_\nu/\partial y^2$ is proportional to $Z_\nu$. The inequality $\sqrt{2 t_e} \ll Y$ means that $Z_\nu$ at $y = \pm Y$ is small in comparison to $Z_\nu(0)$. Finally we conclude that $Z_\nu$ (as well as $Z_{\nu+1}$) depends on y approximately as $\cos(\pi y/2 Y)$. Then the relation (81) yields

$$\gamma \equiv \frac{Z_{\nu+1}}{Z_\nu} = e^{-\frac{\alpha \Delta l}{k_0 T}} \left(1 - \frac{\pi^2 k_0 T \Delta l}{8 \alpha_e Y^2}\right) \tag{82}$$

Hence $Z_\nu$ depends on $\nu$ as $\gamma^\nu$. The specific free energy $\alpha + \delta\alpha$ is equal to $-k_0 T \ln Z_\nu$ divided by a step length $\nu \Delta l$, then

$$\delta\alpha = \varkappa_e \frac{(k_0 T)^2}{\alpha_e Y^2} \tag{83}$$

where $\varkappa_e = \pi^2/8$. The correction, similar to (83), was earlier found for the Kossel model[8]. The obtained value of the numerical factor $\varkappa_e$ is approximate, because the steps, adjacent to a given one, were assumed straight and fixed. The precise analysis of simultaneous fluctuations of many steps for the Kossel model (which is to be published later) leads to Eq. (83), the factor $\varkappa_e$ being equal to $\pi^2/6$. Therefore the last value of $\varkappa_e$ should be used in the general expression (83).

To calculate $\sigma^*_{\lambda\mu}$ it is necessary to define some fixed (with respect to the crystal) axes $x_1$, $x_2$. We shall direct the $x_1$-axis along steps and the $x_2$-axis along the normal **m** (steps on a vicinal surface under consideration are meant). The quantity $\alpha$ in Eq. (29) should be replaced by $\alpha + \delta\alpha$; a step spacing Y and a step tilt angle $\varphi$ (for an arbitrary vicinal surface) should be expressed through the slopes $p_1$, $p_2$. Then the general expression (21) gives

$$\sigma^*_{11} = \frac{\alpha_e}{h\Theta}\,; \qquad \sigma^*_{22} = \frac{(\pi k_0 T)^2}{\alpha_e h^3}\,\Theta \tag{84}$$

The nondiagonal tensor component $\sigma^*_{12}$ is proportional to $\Theta\, d\alpha_e/d\varphi$. Hence $(\sigma^*_{12})^2 \ll \sigma^*_{11}\sigma^*_{22}$ and we can put simply $\sigma^*_{12} = 0$. The stability conditions (14) are fulfilled, thus a vicinal surface is a kind of nonsingular surface, stable against small deviation from the flat average shape. Since $\sigma^*_{22} \ll \sigma^*_{11}$ the resistance to a deviation is the weakest for the cross-section $z x_2$ (such a deviation corresponds to variation of step density along the $x_2$-axis i.e. along **m**).

# 5 Two-Dimensional Nucleation

## 5.1 Critical Nucleus

The edge free energy of a two-dimensional nucleus of the area s reaches its minimum at a certain nucleus shape[21]; the minimum value, $F_e(s)$, is proportional to $\sqrt{s}$. The increase in the total free energy due to the formation of a nucleus is equal to $F = F_e(s) - f q s$ and reaches its maximum at some critical area $s_{cr}$; the maximum increase, $F_{cr}$, equals

$$F_{cr} = \frac{1}{2} F_e(s_{cr}) = f q s_{cr} \tag{85}$$

The critical nucleus corresponds to zero change in F at any small deviations of nucleus boundary (in the linear approximation). Thus the shape of the critical nucleus is given by the equilibrium Eq. (44). The edge free energy $F_e$ is equal to the integral $\oint \alpha\, dl$ over the nucleus boundary. If the integration variable l is changed to the step tilt angle $\varphi$ ($d\varphi = \zeta\, dl$) and the step curvature is expressed from (44), then the final formula for $F_{cr}$ is

$$F_{cr} = \frac{1}{2qf}\int_0^{2\pi} \alpha\,\alpha_e\,d\varphi = \frac{1}{2qf}\int_0^{2\pi}\left[\alpha^2 - \left(\frac{d\alpha}{d\varphi}\right)^2\right]d\varphi \qquad (86)$$

Then the area of the critical nucleus, $s_{cr} = F_{cr}/qf$, is also determined.

## 5.2 Equilibrium Concentration of Nuclei

Let us first define a reference equilibrium contour; it is fixed with respect to the crystal and satisfies Eq. (44). The contour is divided into n elements of length $\Delta l_i$ ($i = 1, \ldots, n$). A nucleus, close to the reference contour (Fig. 8) can be described by the set of normal displacements $\eta_i$ of the elements of nucleus boundary. The equilibrium probability $dP_0$ of finding a nucleus with variables, confined to intervals $d\eta_i$, is equal to the partition function factor dZ corresponding to the nucleus formation. This factor is determined both by the nucleus boundary and by the increase in the number of crystal atoms, i.e. $dP_0$ is given by the expression, similar to (33), if $\Psi$ is replaced by $\Psi_{ef} = \Psi - fqs$

$$dP_0 = \exp\left(-\frac{\Psi_{ef}}{k_0 T}\right)\frac{d\eta_1}{a}\ldots\frac{d\eta_n}{a} \qquad (87)$$

The portion of a boundary, which is nearly parallel to the x-axis, contributes to $\Psi_{ef}$ by the term, defined by Eqs. (37), (42). Particularly, if the step profile $y_i = y(x)$ is smooth, then this contribution is

$$\Psi_{ef} = \nu\psi + \frac{\alpha_e}{2}\int\left(\frac{dy}{dx}\right)^2 dx - qf\int y\,dx \qquad (88)$$

where $\nu$ is the number of step elements in the portion of a nucleus boundary. It is convenient to introduce the deviation $\tilde{y} = y(x) - y_{cr}(x)$ from the reference contour. Then the Eq. (88) will take the following form (after the term, proportional to $d\tilde{y}/dx$ is integrated by parts)

$$\Psi_{ef} = \Psi_{cr} + \frac{\alpha_e}{2}\int\left(\frac{d\tilde{y}}{dx}\right)^2 dx \qquad (89)$$

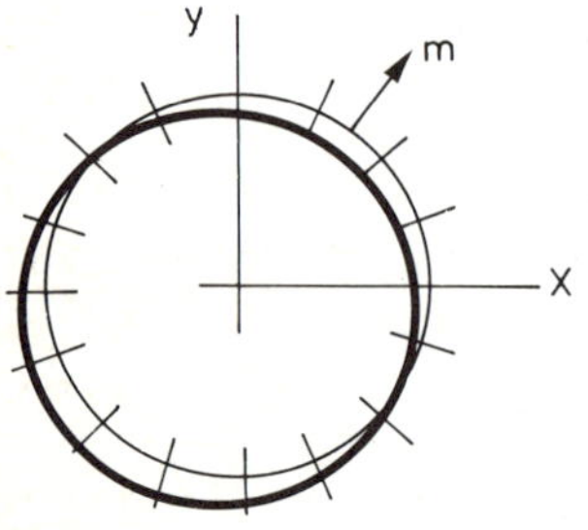


**Fig. 8.** A nucleus (*heavy line*) close to the reference critical contour (*thin line*)

where $\Psi_{cr}$ corresponds to the reference critical profile $y_{cr}(x)$ and equals

$$\Psi_{cr} = \nu\psi - qf\int y_{cr}\,dx \tag{90}$$

because the term, quadratic in $dy_{cr}/dx$, is negligible.

Now $\tilde{y}(x)$ should be expressed through the normal displacement $\eta(l) = \tilde{y}\cos\varphi$ where l is the length along the contour and $\varphi$ is the angle between **m** (the unit normal to the contour) and the y-axis. It follows from the simple relations $d\varphi/dl = \zeta$ and $dx/dl = -\cos\varphi$ that $d\tilde{y}/dx = -d\eta/dl - \zeta\eta\varphi$. This expression is to be substituted into (89). Then the term, linear in $\varphi$, can be written in the form $\varphi\zeta\, d\eta^2/dl$ and integrated by parts; all the terms, quadratic in $\varphi$, can be again neglected. Thus

$$\Psi_{ef} = \Psi_{cr} + \int \frac{\alpha_e}{2}\left[\left(\frac{d\eta}{dl}\right)^2 - \zeta^2\eta^2\right] dl \tag{91}$$

This expression is independent of the special coordinate system x, y used initially. Therefore it holds for any portion of a nucleus boundary, that is, for the whole boundary if $\alpha_e$ and $\zeta$ are considered to be the functions of l or $\varphi$. The quantity $\Psi_{cr}$ for the whole nucleus differs from (90): $\nu\psi$ is to be replaced by the sum $\Sigma\,\psi_i$ over all the elements while $\int y_{cr}\,dx$ should be replaced by the critical area $s_{cr}$. If $\psi_i$ is expressed from (46) and relation (85) is taken into account, then

$$\Psi_{cr} = F_{cr} + \frac{k_0T}{2}\sum_{i=1}^{n} \ln\left(\frac{2\pi k_0 T\,\Delta l_i}{a^2\alpha_{ei}}\right) \tag{92}$$

where $\alpha_{ei} = \alpha_e(\varphi_i)$ and the angle $\varphi_i$ corresponds to the i-th element of the contour. If the integration variable l in Eq. (91) is changed for $\varphi$ and the relation $d\varphi/dl = \zeta = qf/\alpha_e$ is recalled, then

$$\Psi_{ef} = \Psi_{cr} + \frac{qf}{2}\int_0^{2\pi}\left[\left(\frac{d\eta}{d\varphi}\right)^2 - \eta^2\right] d\varphi \tag{93}$$

At this point it is convenient to change the origin of angle $\varphi$: from now on $\varphi$ is considered to be the angle between the normal **m** and the x-axis. The final Eq. (93) for the nucleus free energy $\Psi_{ef}$ is duly invariant under nucleus translation by a vector $\delta\mathbf{r} = (\delta x, \delta y)$, that is, $\Psi_{ef}$ is unchanged if a translational displacement $(\mathbf{m}\cdot\delta\mathbf{r}) = \delta x\cdot\cos\varphi + \delta y\cdot\sin\varphi$ is added to $\eta(\varphi)$.

Coming back to a general state of a nucleus (described by the variables $\eta_i$) one should replace the integral in Eq. (93) by the quadratic form. The resulting expression is simplified if the angle difference between adjacent contour elements, $\Delta\varphi_i = \zeta\Delta l_i$, keeps constant and thus equals $2\pi/n$. Then the element lengths $\Delta l_i$ are specified by the expression

$$\Delta l_i = \frac{2\pi\alpha_{ei}}{qfn} \tag{94}$$

while the expression for $\Psi_{ef}$ takes the form

$$\Psi_{ef} = \Psi_{cr} + \frac{q f n}{4\pi} \sum_{i=1}^{n} [(\eta_i - \eta_{i-1})^2 - \lambda^2 \eta_i^2] \tag{95}$$

where $\lambda = 2\pi/n$. This expression for $\lambda$ can be replaced by

$$\lambda = 2 \sin \frac{\pi}{n} \tag{96}$$

because $n \gg 1$. The aim of this replacement is to keep the invariance of $\Psi_{ef}$ under nucleus translations: $\Psi_{ef}$ is now unchanged if a translational displacement $\eta_i^{(tr)}$ is added to $\eta_i$

$$\eta_i^{(tr)} = \delta x \cdot \cos \varphi_i + \delta y \cdot \sin \varphi_i \tag{97}$$

where $\varphi_i = 2\pi i/n$.

The quadratic form (95) can be written in a general form

$$\Psi_{ef} = \Psi_{cr} + \frac{k_0 T}{2} \sum_{i,k=1}^{n} \phi_{ik} \eta_i \eta_k \tag{98}$$

where the matrix $\phi_{ik}$ is given by

$$\phi_{ik} = \frac{q f n}{2\pi k_0 T} [(2 - \lambda^2) \delta_{ik} - \delta_{i-1\,k} - \delta_{i+1\,k}] \tag{99}$$

Here it is assumed that the Kroneker symbol $\delta_{i'k'}$ adopts the unit value not only at $i' = k'$ but also at $|i' - k'| = n$.

The next step is to express $\eta_i$ through such new variables $\eta_i'$ that the quadratic form (98) is transformed to the combination of squares of $\eta_i'$. For an arbitrary variable transformation

$$\eta_i = \sum_{k=1}^{n} U_{ik} \eta_k' \tag{100}$$

the new matrix $\phi_{ik}'$ is expressed according to (98), (100)

$$\hat{\phi}' = \hat{U}^{\dagger} \hat{\phi} \hat{U} \tag{101}$$

Here the matrices are denoted by the symbol $\wedge$ while the symbol $\dagger$ means transposition. In order to construct the proper transformation $\hat{U}$ we shall first examine the set of eigenvalues of the matrix $\hat{\phi}$. It is easy to check, that the complex vectors $\eta_k^{(m)} = \exp(2\pi jkm/n)$, where j is imaginary unit, are the eigenvectors of $\hat{\phi}$, the eigenvalues $\phi_i$ being equal to

$$\phi_i = \frac{2qfn}{\pi k_0 T}\left[\sin^2\left(\frac{\pi i}{n}\right) - \frac{\lambda^2}{4}\right] \tag{102}$$

where $i = 1, \dots, n$. It follows from (102), (96), that $\phi_1 = \phi_{n-1} = 0$, $\phi_n < 0$ whereas the other eigenvalues are positive. According to the general property of a quadratic form, if $\hat{\phi}$ is transformed to some diagonal matrix $\hat{\phi}'$, then the set of diagonal components $\phi'_{ii} \equiv \phi'_i$ include two zero quantities and one negative quantity, whereas the other ones are positive. The diagonal form of $\hat{\phi}'$ is attained e.g. if the columns of $\hat{U}$ are the mutually orthogonal real eigenvectors of $\hat{\phi}$. Further we shall choose the first and the second columns of $\hat{U}$ in the following form

$$U_{i1} = \cos\frac{2\pi i}{n}; \qquad U_{i2} = \sin\frac{2\pi i}{n} \tag{103}$$

These columns are respectively the real part and the imaginary part of the eigenvector $\eta_i^{(1)}$. According to (97), (100) the new variables $\eta'_1$, $\eta'_2$ coincide with the nucleus translations $\delta x$ and $\delta y$, respectively, hence $\phi'_1 = \phi'_2 = 0$ (because of the translational invariance of $\Psi_{ef}$). The other variables $\eta'_i$ will be ordered in such a manner that $\phi'_3 < 0$ and $\phi'_i > 0$ at $i = 4, \dots, n$. The free energy $\Psi_{ef}$ has a saddle point at $\eta'_i = 0$ $(i \geq 3)$: it decreases with increasing $|\eta'_3|$ and increases with increasing $|\eta'_i|$ at $i \geq 4$. Thus the variable $\eta'_3$ has the sense of a nucleus size whereas $\eta'_4, \dots, \eta'_n$ describe a nucleus shape.

Let us denote by $dN_0$ the average number of nuclei (per unit facet area) having the size and shape parameters between $\eta'_i$ and $\eta'_i + d\eta'_i$ $(i \geq 3)$. If the parameters of a given nucleus are in the intervals $d\eta'_i$ $(i \geq 1)$ then the nucleus center is inside the area $dA = d\eta'_1 d\eta'_2$. Thus the probability $dP_0$ of finding such a nucleus is equal to $dN_0 dA$. Now, if the variables $\eta_i$ in Eq. (87) are expressed through $\eta'_i$, then the nucleus concentration $dN_0$ is obtained

$$dN_0 = a^{-n} \exp\left[-\frac{\Psi_{cr}}{k_0 T} - \frac{1}{2}\sum_{i=3}^{n} \phi'_i \cdot (\eta'_i)^2\right] \cdot |\mathrm{Det}\hat{U}|\, d\eta'_3 \dots d\eta'_n \tag{104}$$

Integrating this expression over the shape parameters $\eta'_4, \dots, \eta'_n$ one finds the concentration of nuclei having the size inside the interval $d\eta'_3$

$$dN_0 = G \cdot |\mathrm{Det}\hat{U}| \cdot (\phi'_4 \dots \phi'_n)^{-1/2} d\eta'_3 \tag{105}$$

where

$$G = a^{-n}(\sqrt{2\pi})^{n-3} \exp\left[-\frac{\Psi_{cr}}{k_0 T} - \frac{1}{2}\phi'_3 \cdot (\eta'_3)^2\right] \tag{106}$$

The product $\phi'_4 \dots \phi'_n$ can be found in the following way. Let us assume for a moment that the number $\lambda$ in Eq. (102) differs from (96) by an infinitesimal amount, that is, $\phi_1 = \phi_{n-1} \neq 0$. Then it follows from (101), (103) that $\phi'_1 = \phi'_2 = \phi_1 n/2$. Now the product $\phi'_1 \dots \phi'_n$ (which is equal to $\mathrm{Det}\,\hat{\phi}'$) is expressed from (101) through $\mathrm{Det}\,\hat{\phi} = \phi_1 \dots \phi_n$ and $\mathrm{Det}\,\hat{U}$. Note that the final expression for $dN_0$ does not contain the unspecified quantity Det $\hat{U}$. According to (102), the product of positive eigenvalues $\phi_i$ includes the factors $[\sin^2(\pi i/n) - \sin^2(\pi/n)]$, $i = 2, \ldots, n-2$. Each factor differs from $\sin^2(\pi i/n)$ by a correction factor which is essential only at i close to 1 or to n and thus equals $(i^2-1)/i^2$ at low i (this expression holds at $i \approx n$ if i is replaced by $n - i$). The total product of correction factors amounts to 1/4 while the product of $\sin^2(\pi i/n)$ can be found by virtue of

the identity $\prod_{i=1}^{n-1} \sin\frac{\pi i}{n} = n\,2^{1-n}$ cited e.g. in Ref. 24.

The quantity $\eta_3'$ characterizes the deviation of a nucleus size from the critical value. It is convenient to introduce the special notation $\xi$ for the size variable (then $\eta_3' = \xi - \xi_{cr}$). Taking into account the relations (106), (92), (94) we can now express the equilibrium nucleus concentration in the final form

$$C_0(\xi) \equiv \frac{dN_0}{d\xi} = \frac{q f}{4\pi^2 k_0 T}(2\pi\phi)^{1/2} \exp\left[-\frac{F_{cr}}{k_0 T} + \frac{\phi}{2}(\xi - \xi_{cr})^2\right] \tag{107}$$

where $\phi = |\phi_3'|$. The nucleation rate (see Sect. 4) does not contain the parameter $\phi$. Thus we need neither to calculate $\phi$ nor to specify the meaning of the size variable $\xi$. Still, for the sake of illustration, we shall consider the particular transformation matrix $\hat{U}$, all the columns of which are the eigenvectors of $\hat{\phi}$. The third column $U_{i3}$ should coincide with the eigenvector $\eta_i^{(n)} \equiv 1$ corresponding to the negative eigenvalue $\phi_n$. It follows from (101), (102) that $\phi_3' = n\phi_n = -2\pi q f/k_0 T$ and $\eta_3' = (\Sigma\,\eta_i)/n$. Thus $\eta_3'$ has the meaning of the normal displacement of the nucleus boundary, averaged over $\varphi$. Particularly, if the step properties are isotropic (i.e. the critical nucleus is a circle) then $\eta_3'$ is proportional to the change in the nucleus area, $s - s_{cr}$, as well as to the change in the number of crystal atoms in the nucleus, $g - g_{cr} = 2\pi r_{cr} q \eta_3'$ (where $r_{cr} = \alpha/qf$ is a critical radius). After $\eta_3' = \xi - \xi_{cr}$ is expressed through g, the formula (107) gives the equilibrium concentration $C_0(g) = dN_0/dg$ of nuclei, containing a given number of atoms. At fixed g the dependence of $C_0(g)$ on f is determined by the factor $\exp(fg/k_0 T)$. If this factor is distinguished, then $C_0(g)$ takes the form which is valid not only at $g \approx g_{cr}(f)$ but at any f, g

$$C_0(g) = \frac{1}{4}\pi^{-3/4}\left(\frac{\alpha}{k_0 T}\right)^{3/2} q^{1/4} g^{-5/4} \exp\left(\frac{fg}{k_0 T} - \frac{2\alpha}{k_0 T}\sqrt{\frac{\pi g}{q}}\right) \tag{108}$$

The pre-exponential factor in (108) depends on g and differs significantly from the commonly adopted constant value q[35, 36], that is, from the number of different postions of a given nucleus per unit facet area. The reason for such a difference is that the free energy of a nucleus boundary differs from $\oint \alpha\, dl$ by some relatively small but absolutely significant correction due to limitation on possible positions of step elements at a given g.

Up to this point we neglected the correction to $\Psi$ due to a step curvature; this correction corresponds to the term $B_e d^2y/dx^2 = -B_e \zeta$ in (35) and results in the additional

factor $\exp\left(\int_0^{2\pi} B_e\, d\varphi\right)$ in the expressions (107), (108). However, this correction is hardly significant (for instance, the convex and the concave step profiles are equivalent in the Kossel model, i.e. $B_e = 0$ in this case).

## 5.3 Nucleation Rate

It is useful to start with the simplest approach to the nucleation problem[36, 37]: a nucleus size $\xi$ (e.g. the average radius, the area, or some other variable) is considered a Markovian quantity that changes at some mean rate $\dot{\xi}(\xi)$; the random change in $\xi$ is characterized by the diffusion coefficient $D(\xi)$. The Einstein relation between $\dot{\xi}(\xi)$ and $D(\xi)$ is

$$\dot{\xi} = \frac{1}{C_0}\frac{d(DC_0)}{d\xi} \approx (\xi - \xi_{cr})\, D\, \phi \tag{109}$$

The second of these expressions follows from (107); the diffusivity $D(\xi)$ is considered a slow function in comparison to $C_0(\xi)$. Differentiating (109) with respect to $\xi$ one obtains

$$\omega \equiv \left.\frac{d\dot{\xi}}{d\xi}\right|_{\xi = \xi_{cr}} = \phi\, D(\xi_{cr}) \tag{110}$$

The quantity $\omega$, defined by this expression, is invariant under the transition from $\xi$ to some other size variable $\xi'(\xi)$. It is an easy task to calculate $\omega$ from the known size dependence of the nucleus growth rate. For instance the radius r of a round nucleus (isotropic case) changes at the mean rate determined by the general Eq. (58): $\dot{r} = v_\infty (1 - r_{cr}/r)$ where $v_\infty = \beta f / k_0 T$ is the mean velocity of a straight step. Putting $\xi = r$ one finds

$$\omega = \frac{v_\infty}{r_{cr}} = \frac{\beta q f^2}{\alpha k_0 T} \tag{111}$$

The opposite case of strong anisotropy was treated in[38]; if there are M equivalent step orientations corresponding to the minimum kink density $\varrho_0$ and hence to the minimum kinetic coefficient $\beta_0$ then

$$\omega = \frac{2}{\pi} \beta_0 q h_e \left(\frac{f}{k_0 T}\right)^2 \tan\frac{\pi}{M} \tag{112}$$

The stationary nucleation rate I (that is, the number of supercritical nuclei arising per unit facet area and per unit time) is determined[36] by the expression which is quite similar to the classical result[39, 40]

$$I = \left(\int \frac{d\xi}{D C_0}\right)^{-1} \tag{113}$$

The main contribution to this integral is due to the vicinity of $\xi_{cr}$ since $C_0(\xi)$ has a sharp minimum at $\xi = \xi_{cr}$ while $D(\xi)$ is a slow function. Substituting Eq. (107) into (113) and expressing $D(\xi_{cr})$ from (110) we obtain the final general formula

$$I = \frac{\omega q f}{4\pi^2 k_0 T} \exp\left(-\frac{F_{cr}}{k_0 T}\right) \tag{114}$$

The stationary nucleation rate is established after some time lag $\tau$. Thus the expression (114) is applicable only if $\tau$ is small in comparison to the new layer expectation time, h/V (where V is the normal growth rate). Different estimates of $\tau$[41–44] do not coincide in details, still each one is the order of the time of diffusion through the near-critical region (in the one-dimensional space of nucleus size), $\tau \sim (\Delta\xi)^2/2D$. Here the half-width of the near-critical region, $\Delta\xi$, corresponds to the ratio $C_0(\xi)/C_0(\xi_{cr}) = e$; according to (107), $\Delta\xi = \sqrt{2/\phi}$. Expressing $\phi$ from (110) one finds $\tau \sim 1/\omega$. Thus the criterion of stationary nucleation has the form $V \ll \omega h$.

Now we proceed to the more rigorous theory of nucleation which deals with simultaneous fluctuations of all n elements of the nucleus boundary. Random displacements $\delta\eta_i$ of variables $\eta_i$ satisfy the condition, similar to (52)

$$\langle \delta\eta_i \delta\eta_k \rangle = 2 D_{ik}\, dt \tag{115}$$

where the diffusion matrix $D_{ik}$ is equal to $D_i \delta_{ik}$. The diffusivity $D_i$ of a given step element i is related to the kinetic coefficient $\beta_i = \beta(\varphi_i)$ according to (57) while the element length $\Delta l_i$ is given by Eq. (94). Thus

$$D_i = \frac{\beta_i f n}{2\pi \alpha_{ei}} \tag{116}$$

The mean velocities $\dot{\eta}_i$ are determined by the Einstein's relations, similar to (109); the concentration $C_0$ should be replaced by the equilibrium distribution function defined by Eqs. (87), (98). Thus

$$\dot{\eta}_i = -\sum_{k=1}^{n} R_{ik} \eta_k \tag{117}$$

where the rate matrix $\hat{R}$ is given by

$$\hat{R} = \hat{D}\hat{\phi} \tag{118}$$

If the variables $\eta_i$ are changed to $\eta_i'$ according to (100) then the new matrix $\hat{\phi}'$ is given by (101) while expressions for $\hat{D}'$ and $\hat{R}'$ follow from (115) and (117) respectively

$$\hat{D}' = \hat{U}^{-1}\hat{D}(\hat{U}^{-1})^{\dagger} \tag{119}$$

$$\hat{R}' = \hat{U}^{-1}\hat{R}\hat{U} \tag{120}$$

The relation (118) keeps valid i.e. $\hat{R}' = \hat{D}'\hat{\phi}'$.

There exists such a transformation $\hat{U}$ that both $\hat{D}'$ and $\hat{\phi}'$ take a diagonal form (then $\hat{R}'$ is diagonal too). A transformation of this kind can be constructed, e.g. in two stages: first $D_{ik}$ is reduced to unit matrix $\delta_{ik}$ by virtue of scaling transformation $\eta_i = \eta_i' \sqrt{D_i}$; then $\hat{\phi}'$ is reduced to a diagonal matrix by virtue of a unitary transformation ($\hat{U}^{-1} = \hat{U}^{\dagger}$) while

$\hat{D}'$ retains its unit form. It follows from the diagonal form of $\hat{R}'$ and from the relation (120) that each column of $\hat{U}$ is an eigenvector of $\hat{R}$. Hence the first and the second columns of $\hat{U}$ can be chosen in the previous form (103) because these columns are the eigenvectors of $\hat{\phi}$ corresponding to zero eigenvalues (then they are also the eigenvectors of $\hat{R} = \hat{D}\hat{\phi}$ corresponding to zero eigenvalues). Thus the variables $\eta_1'$, $\eta_2'$ are again the nucleus translations. The variables $\eta_i'$ are assumed to be ordered in the same manner as before: $\eta_3'$ describes a nucleus size ($\phi_3' < 0$) whereas $\eta_4', \ldots, \eta_n'$ describe a nucleus shape ($\phi_i' > 0$).

The diagonal form of both $\hat{R}'$ and $\hat{D}'$ means that both average and random change in any variable $\eta_i'$ proceeds independently of the other variables $\eta_k'$. Thus variation of the size variable $\eta_3'$ is essentially the one-dimensional drift and diffusion considered above, while the distribution of $\eta_4', \ldots, \eta_n'$ is the equilibrium. The kinetic factor $\omega$, that determines the nucleation rate (114), is equal to $-R_3'$ according to the rate equation $\dot{\eta}_3' = -R_3'\eta_3'$ and the relation (110). The quantity $R_3'$ is the sole negative eigenvalue of $\hat{R}'$ since the sign of an eigenvalue $R_i' = D_i'\phi_i'$ coincides with that of $\phi_i'$. The general transformation (120) is known to have no effect on the eigenvalues i.e. the set of $R_i'$ coincides with the set of eigenvalues $R_i$ of matrix $\hat{R}$. Thus the factor $\omega$ is equal to the sole positive eigenvalue of matrix $-\hat{R} = -\hat{D}\hat{\phi}$. If $\phi_{ik}$ and $D_{ik} = D_i\delta_{ik}$ are substituted from (99) and (116) respectively, and $\eta_i$ is considered a smooth function of the angle $\varphi = 2\pi i/n$ then $\omega$ can be determined by solving the equation

$$\frac{d^2\eta}{d\varphi^2} + \eta = \frac{\omega\,\alpha_e\,k_0 T}{\beta\, q\, f^2}\,\eta \tag{121}$$

For isotropic step Eq. (121) has a solution $\eta = \text{const}$; the corresponding value of $\omega$ coincides with Eq. (111).

In the opposite case of strong anisotropy the function $\alpha_e/\beta$ has a sharp maximum at $\varphi = 0$ (and at the other $M - 1$ equivalent step orientations). The behavior of $\alpha_e/\beta$ in the vicinity of each maximum is described by Eqs. (73), (74), (71). Multiplying Eq. (121) by $\cos\varphi$ and integrating between 0 and $\pi/M$, and then repeating this operation with $\cos\varphi$ replaced by $\sin\varphi$, we obtain the system of two homogeneous equations for $\eta(0)$ and $\eta(\pi/M)$. The resulting value of $\omega$ coincides with Eq. (112).

In conclusion it is useful to write down the explicit expression for the nucleation rate in the isotropic case. It follows from (114), (111) and (86) that

$$I = \frac{\beta\, q^2 f^3}{4\,\alpha\,(\pi\, k_0 T)^2}\ \exp\left(-\frac{\pi\,\alpha^2}{f\, q\, k_0 T}\right) \tag{122}$$

## 5.4 Growth Rate

If the facet is free from dislocations and other sources of steps then the normal growth rate V is determined by the nucleation rate I, the tangential velocity of steps v (which is close to $v_\infty$) and the size $r_n$ of a nucleation zone. If the facet is isothermal then $r_n$ coincides with the facet size. However, a facet is usually nonisothermal (Fig. 5); nucleation occurs mainly in a small zone where the supercooling is close to its maximum value $\Delta T_{max}$. The nucleation zone size $r_n$ can be defined by equation $I/I_{max} = 1/e$ where $I_{max}$ is the maximum nucleation rate (at $\Delta T = \Delta T_{max}$). For instance if the temperature distribution along the facet (Fig. 5 a) is parabolic ($\Delta T = \Delta T_{max}(1 - r^2/r_0^2)$ where $r_0$ is the facet radius) then it follows from (122), (6)

$$r_n = r_0 (k_0 T/F_{cr})^{1/2} \tag{123}$$

where $F_{cr}$ corresponds to the facet centre.

Within the expectation time of a new layer, $\sim h/V$, the island of a previous layer expands to the distance $\sim vh/V$. If this distance is greater than $r_n$, then V/h (the number of new layers arising per unit time) is equal to the integral of I over the nucleation zone, i.e. $V \approx \pi r_n^2 h I_{max}$. However, it is likely that the opposite inequality $vh/V \ll r_n$ is fulfilled, that is, the nucleus of a new layer arises on a relatively small island[22, 35]. In this case V/h does not depend on $r_n$; it is proportional to $(Iv^2)^{1/3}$ according to dimensional considerations. The proportionality factor (of order unity) was estimated in Refs.[35, 45–47], yet this factor is hardly essential since V/h contains the exponential factor. The final expression for growth rate (in the case of isotropic step) follows from (56), (122)

$$V = \beta h \left(\frac{q^2 k_0 T}{4\pi^2 \alpha}\right)^{1/3} \left(\frac{f_{max}}{k_0 T}\right)^{5/3} \exp\left(-\frac{\pi \alpha^2}{3 q k_0 T f_{max}}\right) \tag{124}$$

where $f_{max}$ corresponds to the maximum supercooling $\Delta T_{max}$. The formula (124) is based on the refined expression (107) for the equilibrium nucleus concentration. So it is somewhat different from the result obtained in Refs. 22, 35.

If a dislocation emerges at some point $\mathscr{D}$ of a facet, and the Burger's vector of the dislocation is not parallel to the facet, then a step goes out from the point $\mathscr{D}$. At $f \neq 0$ the step takes a spiral form[4, 21] characterized by the step spacing $Y = \varkappa_d r_{cr}$ at large distance from $\mathscr{D}$; the numerical factor $\varkappa_d$ is equal to $\approx 19$ in the isotropic case. Thus the growth rate, governed by the spiral source of steps, is

$$V = \frac{h v_\infty}{\varkappa_d r_{cr}} = \frac{\beta h q}{\varkappa_d \alpha k_0 T} f^2 \tag{125}$$

where the driving force f corresponds to the point $\mathscr{D}$. If there is only one such point $\mathscr{D}$ on the facet then a change in location of $\mathscr{D}$ leads to a variation of the growth rate V and, consequently, of the facet size and even of the external crystal shape if the facet adjoins the three-phase line $\mathscr{L}$ (Fig. 5 b). Such an effect was indeed observed when a silicon crystal contained one or more dislocations[48]. On the other hand, if there are many dislocation points $\mathscr{D}$ on the facet, then the growth rate is determined by the spiral source of the maximum supercooling, i.e. one should substitute $f \approx f_{max}$ into Eq. (125).

The expressions (124), (125) can be used to determine the basic step parameters $\alpha$ and $\beta$ if the maximum supercooling $\Delta T_{max}$ is known both for dislocation-free and dislocated crystals. The quantity $\Delta T_{max}$ can be either directly measured[49, 50] or calculated[22] from the gap between the facet and the extrapolated solidification isotherm $T_0$ as revealed by crystal decantation or by impurity striations in the longitudinal crystal section. This method was used[22] to find $\alpha$ and $\beta$ for steps on {111}-facet of silicon crystal grown from the melt. The expression for the nucleation – limited growth rate, used in[22] was similar to that of Ref. 35. If the refined expression (124) is used then one obtains slightly different numbers: $\alpha \approx 2 \times 10^{-6}$ erg/cm and $\beta \approx 2 \times 10^4$ cm/s ($\beta_T \approx$ 50 cm/s °C).

## 6 Summary

Singular and vicinal surfaces are the most important objects of the theory because crystal growth at these surfaces requires considerable deviation from the conditions of phase equilibrium. In this case, the growth processes are reduced to generation and motion of steps. A step can be treated as a system of small fluctuating elements, the element length $\Delta l$ being much greater than the atomic spacing. The final results are independent of $\Delta l$; they contain only the specific free energy $\alpha$ and the kinetic coefficient $\beta$ of steps. Although $\alpha$ and $\beta$ generally depend on a step orientation angle $\varphi$, the anistropy is likely to be small in the case of the crystal-melt interface since the minimum kink density $\varrho_0$ is rather high[21, 28]. Then all the properties of the interface (the nucleation rate (122), the dependence of growth rate on supercooling (124), (125), the stability tensor (84) of a vicinal surface) are expressed through the two basic constants $\alpha$ and $\beta$ which can be determined from experimental data. Any atomic model of a surface would involve more than two basic parameters (at least one energy parameter and a set of attachment probabilities for different sites on a surface). If the step properties are essentially anisotropic (i.e. the minimum kink density $\varrho_0$ is low) then $\alpha(\varphi)$, $\beta(\varphi)$ are determined by the three basic parameters $\alpha_0$, $\beta_0$, $\varrho_0$ which correspond to the step, parallel to close-packed atomic rows.

It is necessary to underline that the approach developed in the present paper is justified only if the characteristic surface scale (such as the critical radius $r_{cr}$ or the step spacing Y) is large in comparison both to the correlation length $l_c$ and the kink spacing $1/\varrho_0$. This condition is fulfilled for a facet at sufficiently low growth rate V.

Although only pure crystals were considered above, the same conception of fluctuating steps can be applied to the problem of impurity incorporation[26] and of impurity effect on growth rate[29]. Further progress may be associated with generalization of the theory to the case of solid solutions. It is also interesting to apply the method described in Sect. 4 to heterogeneous two-dimensional nucleation (at the intersection of a facet and some other interface) and to three-dimensional nucleation.

## 7 References

1. Gibbs, J. W.: The Collected Works, London, 1928
2. Landau, L. D., Lifshitz, E. M.: Statistical Physics, Oxford, Pergamon 1958
3. Davies, G. R.: Metallurg. Rev. *10*, 173 (1965)
4. Cabrera, N., Levine, M. M.: Phil. Mag. *1*, 450 (1956)

5. Gliksman, M. E., Vold, C. L.: Surf. Sci. *31*, 50 (1972)
6. Cabrera, N.: Disc. Faraday Soc. *28*, 16 (1959)
7. Chernov, A. A.: Soviet Phys.-Usp. *4*, 116 (1961)
8. Voronkov, V. V.: Soviet Phys.-Cryst. *12*, 728 (1967)
9. Herring, C.: Surface Tension as a Motivation for Sintering, in: The Physics of Powder Metallurgy, p. 143, New York – Paris, Mc Grow Hill 1951
10. Chui, S. T., Weeks, J. D.: Phys. Rev. *B 14*, 4978 (1976)
11. van Beijeren, H.: Phys. Rev. Lett. *38*, 993 (1977)
12. Leamy, H. J., Gilmer, G. H., Jackson, K. A.: Statistical Thermodynamics of Clean Surfaces, in: Surface Physics of Materials, v. 1 (Blakely, J. M., ed.) p. 121, New York – San Francisco, Acad. Press 1975
13. Temkin, D. E.: On Molecular Roughness of Crystal – Melt Interface, in: Crystallization Processes, p. 15, New York, Consultants Bureau 1966
14. Cahn, J. W.: Acta Met. *8*, 554 (1960)
15. Cahn, J. W., Hillig, W. B., Sears, G. W.: ibid. *12*, 1421 (1964)
16. Bolling, G. F., Tiller, W. A.: J. Appl. Phys. *31*, 1345 (1960)
17. Voronkov, V. V.: Soviet Phys.-Cryst. *19*, 573 (1974)
18. Voronkov, V. V.: ibid. *23*, 137 (1978)
19. Voronkov, V. V.: J. Crystal Growth *52*, 311 (1981)
20. Frenkel, J.: J. of Phys. (USSR) *9*, 392 (1945)
21. Burton, W. K., Cabrera, N., Frank, F. C.: Phil. Trans, Roy. Soc. (London) *A 243*, 299 (1951)
22. Voronkov, V. V.: Soviet Phys.-Cryst. *17*, 807 (1972)
23. Müller-Krumbhaar, H., Burkhardt, T. W., Kroll, D. M.: J. Crystal Growth *38*, 13 (1977)
24. Gradstein, I. S., Rizhik, I. M.: Tables of Integrals, Sums, Series and Products, Moscow, Nauka 1971 (in russ.)
25. Carslaw, H. S., Jaeger, J. C.: Conduction of Heat in Solids, Oxford, Clarendon 1959
26. Voronkov, V. V.: Dope Uptake Factor in Relation to Growth Rate and Surface Inclination, in: Growth of Crystals, v. 11 (Chernov A. A., ed.) p. 364. New York – London, Consultants Bureau 1979
27. Voronkov, V. V.: Soviet Phys.-Cryst. *18*, 19 (1973)
28. Voronkov, V. V.: ibid. *15*, 8 (1970)
29. Voronkov, V. V.: ibid. *19*, 296 (1974)
30. Lifshitz, I. M., Chernov, A. A.: ibid. *4*, 747 (1959)
31. Chernov, A. A., Dukova, E. D.: ibid. *5*, 627 (1960)
32. Mullins, W. W., Hirth, J. P.: J. Phys. Chem. Solids *24*, 1391 (1963)
33. Schwoebel, R. L.: J. Appl. Phys. *40*, 614 (1969)
34. Bennema, P., Van Rosmalen, R.: Simulation of Modes of Vibrations in Trains of Steps, in: Growth of Crystals, v. 11 (Chernov A, A., ed.) p. 162, New York – London, Consultants Bureau 1979
35. Hillig, W. B.: Acta Met. *14*, 1868 (1966)
36. Voronkov, V. V.: Soviet Phys.-Cryst. *15*, 979 (1970); A functional approach to nucleation was given by J. S. Langer, Ann. Phys. (N.Y.) *54*, 258 (1969)
37. Stoyanov, S.: Surf. Sci. *15*, 558 (1969)
38. Voronkov, V. V.: The Rate of Two-dimensional Nucleation in case of Anisotropic Step Kinetics, in: Rost i Legirovanie Poluprovodnikovih Kristallov i Plenok (Growth and Doping of Semiconductor Crystals and Films) part 2 (Alexandrov, L. N., ed.) p. 46, Novosibirsk, Nauka 1977 (in russ.)
39. Zeldovich, J. B.: J. Exper. Theor. Phys. (USSR) *12*, 525 (1942)
40. Frenkel, J. I.: Kineticheskaya Teoriya Zhidkostey (Kinetic Theory of Liquids) Leningrad, Nauka 1975 (in russ.)
41. Ljubov, B. J., Roitburd, A. L.: Dokl. Acad. Nauk (USSR) *111*, 630 (1956)
42. Wakeshima, H.: J. Phys. Soc. Japan *10*, 374 (1955)
43. Kashchiev, D.: Surf. Sci. *14*, 209 (1969)
44. Binder, K., Stauffer, D.: Adv. Phys. *25*, 343 (1976)
45. Borovinskiy, L. A., Tzindergozen, A. N.: On a Perfect Crystal Face Growth in case of Many Nuclei in each Layer, in: Uchen. Zapiski (Proceedings) of Leningrad State Teach. Inst., vol. 431, p. 27, Novgorod 1969 (in russ.)

46. Alexandrov, L. N., Entin, I. A.: Soviet Phys.-Cryst. *20*, 694 (1975)
47. Belenkiy, V. Z., Ljubitov, Yu. N.: ibid. *23*, 705 (1978)
48. Voronkov, V. V., Pankov, V. M.: ibid. *20*, 697 (1975)
49. Brice, J. C., Whiffin, P. A. C.: Solid State Electr. *7*, 183 (1964)
50. Pogodin, A. I., Eidenzon, A. M.: The Influence of Dislocations and Impurities on the Local Supercooling at Crystallization Front of Melt-Grown Silicon Crystals of Different Orientations, in: Protzessi Rosta Puluprovodnikovih Kristallov i Plenok (Processes of Growth of Semiconductor Crystals and Films) (Kuznetzov, F. A., ed.) p. 170, Novosibirsk, Nauka 1981 (in russ.)

# Surface and Volume Diffusion Controlling Step Movement

**J. P. van der Eerden**

RIM Laboratory of Solid State Chemistry, Toernooiveld, 6525 ED Nijmegen, The Netherlands

*An, often rate determining, step in the growth of crystals is the transport of growth units from the bulk mother phase to the steps, where they can be incorporared in the crystal. When convection dimishes rapidly upon approaching the crystal surface, then a diffusion boundary layer can be defined. Growth units may reach the steps either by diffusing within this layer directly or indirectly, after surface diffusion. In this article the coupled diffusion problem is, for an arbitrary step pattern, translated into an equation of motion for steps, using Green's function methods. The effective interaction between moving steps, as induced by the associated diffusion fields, is described and evaluated for large and small distances. Solutions of the equation of motion for simple step geometries, and approximations for general patterns are reviewed.*

*The significance of this approach lies in the applicability to general step patterns. Indeed, even for the rather complicated step patterns which are encountered in experimental observations, the diffusion problem can be approximately solved. This approximate solution can help to determine the relevant crystal growth parameters, and to predict crystal growth rates. In future numerical and analytical approximations of both the diffusional interaction itself and the equation of motion have to be developed. This may, in turn, enable an interpretation of experimentally observed step patterns, which enlarges our understanding of the crystal growth process.*

# 1 Introduction

One of the aspects of crystal growth is the transport of growth units. They are furnished from what we shall call the mother phase. In many experimental situations the mother phase will be well mixed due to convection, turbulence etc. Close to the crystal surface, however, the flow is retarded and the contribution of diffusion to the transport of growth units increases relative to the convective contribution. If the growth units reach the crystal surface they can be incorporated directly into the steps; the direct incorporation or Chernov mechanism[1]. Alternatively, however, they can be deposited on the relatively smooth regions between steps and reach the steps after surface diffusion: the indirect incorporation or Gilmer, Ghez and Cabrera[2-4] mechanism.

In this chapter we assume that a homogeneous mother phase can be separated from a diffusion boundary layer with a constant thickness $\delta$. Of course, strictly spoken, this can not be true. Both mother phase and boundary layer constitute one region in which the Navier-Stokes equations are satisfied. These equations, however, can be solved in some special cases only, the infinite rotating disk with uniformly absorbing surface being one of these[5]. In such a solution the concentration of growth units increases smoothly with the distance from the surface of the disc. It follows that if this exact solution is replaced by a linear variation until a distance $\delta$, then $\delta$ is given by[5]

$$\delta \approx 1.61\,(D/\nu)^{1/3}(\nu/\omega)^{1/2} \approx 0.5\,(D/\nu)^{1/3}\delta_0\,, \tag{1}$$

where $D$ is the diffusion constant of the growth units in the solution, $\nu$ the kinematic viscosity, $\omega$ the rotation frequency of the disc and $\delta_0$ the hydrodynamic boundary layer thickness, i.e. the distance at which the solution flows with 80% of the main stream velocity towards the disc. It is the object of this paper to investigate the flux of growth units towards a crystal surface which does not absorb uniformly but mainly at step positions. In that case the convective diffusion problem can not be solved any more. If it could be solved it would give rise to a certain net flux of growth units, which could have been generated as well from a pure diffusion region (convection neglected) in contact with a homogeneous mother phase. Its thickness $\delta$ would still roughly be given by Eq. (1). Accordingly we study in this paper non-convective diffusion in a boundary layer whose thickness $\delta$ is a free parameter of the model.

We now introduce the other parameters of our transport model, which is depicted schematically in Fig. 1. The flux of growth units is directed partly towards a step strip of width $a$ along all steps on the crystal surface, and coupled to the advance rate with a characteristic length $\Lambda_c$. The rest of the flux is towards step free regions and coupled with a length $\Lambda$ to the local concentration of adsorbed isolated growth units. They return after a mean surface diffusion free path $\lambda$ to the diffusion layer or they reach the step strip and lead to step movement with a characteristic length $\Lambda_s$. The parameters and their physical meaning are given in Table 1. Their precise role will become clear in Sect. 2.5.

In this way we have formulated a coupled two and three dimensional diffusion problem in terms of the six parameters, $\delta$, $a$, $\Lambda_c$, $\Lambda$, $\Lambda_s$ and $\lambda$. In general we have reasonable independent estimates for $\delta$ (see Eq. (1)) and $a$ (the step height) only. For the other parameters estimations on the basis of activation energies[4,6] depend critically on the exact value of these energies and may be orders of magnitude wrong[7]. They have to be

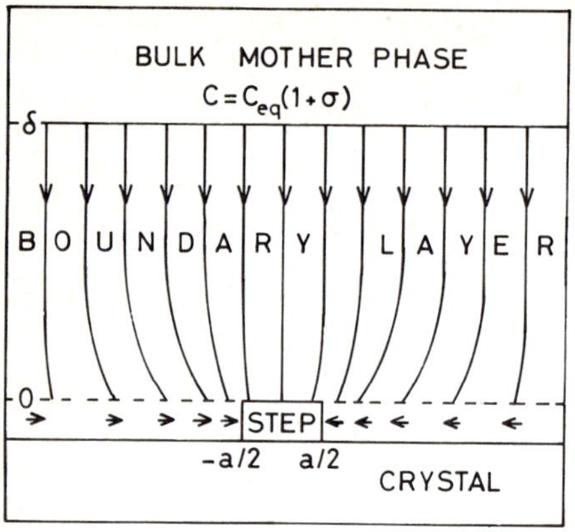


**Fig. 1.** The general model. There is a net flux of growth units from a homogeneous mother phase with concentration $C_\infty = C_{eq}(1 + \sigma)$ through a diffusion boundary layer to the crystal. Some are directly incorporated in a step strip **S** of width *a*, others diffuse over the surface to the step

**Table 1.** Characteristic length scales

| | |
|---|---|
| $\delta$ | boundary layer thickness |
| $\Lambda$ | impedance for volume-surface incorporation |
| $\Lambda_c$ | impedance for volume-step incorporation |
| $\Lambda_s$ | impedance for surface-step incorporation |
| $\lambda$ | surface mean free path of growth units |
| $a$ | width of step strip |

found from quantitative interpretation of experimental data. It turns out that experiments on slightly misoriented faces[8–10] can be interpreted with the results of Refs. (1, 2) if either the indirect or the direct incorporation mechanism is omitted, or with the general results given in Ref. (11) and in Sects. (2.4) and (3.3). This interpretation, in principle, provides the necessary parameter values.

In order to keep the problem feasible the gradients along the steps due to the exact location of kink sites[4] are neglected, and the steps are considered as continuous smoothly curved lines. Also diffusion along the steps[12] is not taken into account. The stationary diffusion model[4] is used.

The diffusion problem in the diffusion layer leads to a three dimensional Laplace equation and the surface diffusion to a two dimensional Helmholtz equation. Using the Green's function for the diffusion layer and the potential function for the surface these coupled partial differential equations, combined with the boundary conditions, can be transformed[13] into two coupled one dimensional integral equations[11] for the direct and indirect contribution to the step velocity.

The chapter is built up as follows. After this introduction in Sect. 2 the physical ideas and mathematical foundation of the equation of motion are sketched and exact solutions are given for simple, yet important, cases. In Sect. 3 the concept of an effective diffusional interaction is described and the strength of this interaction is evaluated for several typical step patterns. In Sect. 4 limiting cases are given in which the interactions, and hence the equation of motion as well simplify. In Sect. 5 an approximate solution method for general step patterns is given and applied to competitive nucleation between steps of a misoriented surface. The problem of an archimedean growth spiral is discussed finally.

# 2 The Equation of Motion

## 2.1 Formulation and Interpretation

In Sect. 2.5 we shall sketch how the following integral equation of motion can be derived:

$$v(s_0) + \int K(s_0, s)v(s)\mathrm{d}s = v_\mathrm{m}(s_0) \ . \tag{2}$$

Physically the form of the equation can be understood as follows.

$v(s_0)$ is the advance velocity of the step in the direction of the normal to the step at a point $\mathbf{r}(s_0)$ of the step which is parametrized by the arc length $s_0$. When there would be no interaction through the various diffusion fields the step would move with some maximal velocity $v_\mathrm{m}(s_0)$. In fact, however, $v(s_0)$ is modified by other parts of the step pattern, which move simultaneously. Step movement does not take place without transport of growth units and it modifies, therefore, the diffusion field in its environment. This modification, in turn, modifies the advance rate of steps in this environment and we end up with an effective interaction between step velocities which is described by the kernel function $K(s_0, s)$.

This kernel depends on the distance

$$\Delta = |\mathbf{r}(s) - \mathbf{r}(s_0)| \tag{3}$$

between the points of the step pattern with arc length $s_0$ and $s$ and it decays exponentially fast at large distances. Explicit expressions for $K$ and $v_\mathrm{m}$ will be given in Sect. 2.5 and 3.1.

It is convenient to write Eq. (2) and similar equations which will be derived below in the symbolical form:

$$\mathbf{v} + \mathbf{K}\mathbf{v} = \mathbf{v}_\mathrm{m} \ . \tag{4}$$

Here $\mathbf{v}$ and $\mathbf{v}_\mathrm{m}$ are real functions defined on the steps and the operator $\mathbf{K}$ transforms the function $\mathbf{v}$, in the usual way, into the new function $\mathbf{Kv}$ given by

$$(\mathbf{Kv})(\mathrm{s}_0) = \int \mathbf{K}(\mathrm{s}_0, \mathrm{s})\mathbf{v}(\mathrm{s})\mathrm{ds} \ , \tag{5}$$

the integration being along all steps which are present on the surface and in such a direction that the terrace at the right hand side is always lower than the one at the left hand side.

It should be mentioned that the diffusion problem is slightly more complicated than mentioned until now. It turns out that one has to distinguish two contributions to the step velocity. First, the direct incorporation from the boundary layer (the volume) into the step, and, second, the indirect incorporation via the surface. These two contributions can be grouped in a two dimensional vector $\mathbf{v}$ and Eq. (4) still holds when the kernel $\mathbf{K}$ is conceived as a two dimensional operator.

## 2.2 Analogy with Electrostatics

In order to get some feeling for the physical meaning of the integral equation Eq. (1) let us consider an electrostatical analogon. Then the role of concentration of growth units is played by the electrical potential $\Phi$ and the net flux $D\nabla C$ is replaced by the electrical field $\mathbf{E} = \nabla\Phi$. The steps are the analogons of metallic conducting strips, mounted on an infinite flat surface, and the supersaturation of the mother phase produces an external field. Velocities of steps correspond to net electrical forces normal to the conductor surface, and the maximal velocity $\mathbf{v}_m$ to the force which would be felt if the external electrical field would not be modified by the presence of induced charges on the conductor. The kernel is the electrical potential of a unit point charge. In order to have a complete formulation of the problem we need boundary conditions. Both in the electrical and in the diffusional case we assume that the boundary (conductor or step) has such physical properties that (i) the potential is continuous (i.e. only charges, no dipoles on the conductor), and (ii) there is a linear relation between the potential and its normal derivative (electrical force or advance velocity).

In the electrical case one expects that the induced charge distribution produces an electrical potential outside the conductor. The superposition principle learns that this potential can be written as the sum of potentials of unit point charges which are distributed over the conductor according to the actual charge density. On the other hand one of Maxwell's equations tells us that this charge density is proportional to the divergence of the electrical field and hence in this case to the net force exerted on the surface of the conductor. At this point we found that the electrical potential around a charged conductor is given by an expression which is essentially the same as the second term of Eq. (2). The last step is to observe that this expression remains valid when one approaches the conductor and to apply the boundary condition stated above. Translating this back to the diffusion problem we see that the integral in Eq. (2) is proportional to the growth unit concentration due to all absorbing steps on the crystal surface. This growth unit concentration is itself proportional to the step velocity and this leads to the first term and the right hand side of this equation.

As a matter of fact the situation is somewhat more complicated. In the boundary layer the concentration satisfies Laplace's equation, just as the electrical potential, since it expresses conservation of particles. On the surface, however, there are exchange processes with the boundary layer, such that Laplace's equation is modified. Moreover, the particles arrive at the step from two different processes, direct integration from the boundary layer, or indirect integration via the surface. This means that Eq. (2) actually is replaced by a set of two coupled equations and that the kernels are more complicated than the $r^{-1}$ and $\ln r$ potentials of three and two dimensional electrostatics. The velocity contribution due to direct integration has to be compared with an electrical force, perpendicular to the surface on which the conductor strips are mounted. The indirect contribution corresponds to the lateral forces which tend to split the strips.

## 2.3 Exact Solution for a Single Infinite Straight Step

In the case of a straight step it is relatively easy to obtain an exact solution of the equation of motion. This solution is found by Fourier transforming Eq. (2). Indeed, if the step is situated along the X-axis, we may define the Fourier transform of v(x) by:

$$\tilde{\mathbf{v}}(k) \equiv \frac{1}{2\pi} \int_{-\infty}^{\infty} e^{ikx} \mathbf{v}(x) dx . \tag{6}$$

Note that the distance $\Delta$ on which the kernel **K** depends is, by Eq. (3), for two positions $s = x$ and $s' = x'$ on the step given by $|x - x'|$. Therefore, upon multiplying Eq. (1) by $\exp(ikx)/2\pi$ and integrating $x$ from $-\infty$ to $\infty$ we get

$$\tilde{\mathbf{v}}(k) + 2\pi\tilde{\mathbf{K}}(k)\tilde{\mathbf{v}}(k) = \tilde{\mathbf{v}}_m(k) \tag{7}$$

where $\tilde{\mathbf{K}}$ is the Fourier transform of $\mathbf{K} = \mathbf{K}(|x - x'|)$. Using the inversion formula we obtain the solution

$$\mathbf{v}(x) = \int_{-\infty}^{\infty} [1 + 2\pi\tilde{\mathbf{K}}(k)]^{-1} \tilde{\mathbf{v}}_m(k) e^{-ikx} dk \tag{8}$$

To be more specific let us consider two special cases. First the trivial case in which the maximal velocity is constant along the step. Then $\tilde{\mathbf{v}}_m(x) = \mathbf{v}_m$ and hence $\mathbf{v}_m(k) = \mathbf{v}_m \delta(k)$. Then Eq. (8) can be combined with Eq. (6) for $\tilde{\mathbf{K}}(0)$ to give the velocity of an isolated straight step:

$$\mathbf{v}(x) = [1 + \int_{-\infty}^{\infty} \mathbf{K}(|u|) du]^{-1} \mathbf{v}_m \tag{9}$$

This result could be expected since if $\mathbf{v}_m$ does not depend on the step position then $\mathbf{v}(x)$ will not either. But then also $\mathbf{v}(x') = \mathbf{v}(x)$ and this constant can be taken out of the integral in Eq. (2) and Eq. (9) is then found immediately.

The second example is less trivial. Assume that $\mathbf{v}_m(x)$ is not a constant but is reduced in a region around $x = 0$. Physically this occurs, e.g., if the step approaches a region of stress. Of course, this stress will locally reduce the actual velocity and the step will not remain straight. Neglecting this deformation we see that the step velocity is given by Eq. (8). As an approximation we take

$$\mathbf{v}_m(x) = \mathbf{v}_m[1 - \varepsilon \exp\{-(x/d)^2\}] . \tag{10}$$

The Fourier transform of the maximal velocity is

$$\tilde{\mathbf{v}}_m(k) = \mathbf{v}_m \left[ \delta(k) - \frac{\varepsilon d}{2\sqrt{\pi}} \exp\left\{ - \left( \frac{kd}{2} \right)^2 \right\} \right] \tag{11}$$

Hence we find from Eq. (8) after some rearrangement.

$$\mathbf{v}(x) = \left\{ [1 + 2\pi\tilde{\mathbf{K}}(0)]^{-1} - \frac{\varepsilon d}{2\sqrt{\pi}} \exp\left\{ - \left( \frac{x}{d} \right)^2 \right\} \int_{-\infty}^{\infty} \exp\left\{ - \left( \frac{kd}{2} + \frac{ix}{d} \right)^2 \right\} \right.$$

$$\left. [1 + 2\pi\tilde{\mathbf{K}}(k)]^{-1} dk \right\} \mathbf{v}_m \tag{12}$$

The first term represents the unperturbed situation. If the interaction range is very small then $\tilde{\mathbf{K}}$ will decay very slowly with $|k|$ and the relative retardation of the step equals the relative decrease of the maximal velocity since $\mathbf{K}(k)$ can be replaced by $\mathbf{K}(0)$ in the second term.

It may be useful to remark that, although the solution, Eq. (8), is interpreted as one dimensional it can be used as well when $\mathbf{v}$ is the two dimensional vector, built up from direct and indirect contributions.

## 2.4 Exact Solution in the Case of Uniform Velocities

There are cases where the symmetry of the problem shows that the velocities of the steps are constant along the steps. This occurs in the absence of stress for a single, infinite, straight step, for an infinite equidistant sequence of such steps and for a single circular step. In such a case the integral equation reduces to a linear equation for the step velocity, which can easily be solved.

We have to distinguish two different contributions, $v_v$ for direct incorporation of growth units from the volume of the boundary layer, and $v_s$ for indirect incorporation via the surface. The velocities $\mathbf{v}$ and $\mathbf{v}_m$ in Eq. (4) should be considered as two dimensional vectors and the kernel $\mathbf{K}$ as a two dimensional operator with components $K^{ss}$, $K^{sv}$, $K^{vs}$ and $K^{vv}$. We give the solution in the general case.

Throughout the text we will denote the integral of $K^{\alpha\beta}$ (where $\alpha$ and $\beta$ are $v$ or $s$) along the complete step pattern by $I^{\alpha\beta}$, sometimes with a symbolical subscript to indicate the step pattern under consideration:

$$I^{\alpha\beta}(s) = \int \mathbf{K}^{\alpha\beta}(s, s')ds' \; . \tag{13}$$

For the present problem the integral does not depend on the fixed step position $\mathbf{r}(s)$ for which it is evaluated, and the argument $(s)$ can be omitted. Substituting Eq. (13) in Eq. (4) we find the solution $v = v_s + v_v$ from

$$Iv_v = v_{mv}(1 + I^{ss}) - v_{ms}I^{vs} \tag{14}$$

$$Iv_s = v_{ms}(1 + I^{vv}) - v_{mv}I^{sv} \tag{15}$$

$$I = (1 + I^{vv})(1 + I^{ss}) - I^{vs}I^{sv} \tag{16}$$

In the sequel we give expressions for $I^{\alpha\beta}$ in several cases.

## 2.5 Sketch of the Derivation

The integral equation of motion which has been presented in the Sect. 2.1 can be seen as an alternative formulation of two coupled diffusion problems. As usual, the diffusion problem naturally leads to a set of differential equations supplemented with boundary conditions.

Using Green's function technique it is possible to derive the integral equation, Eqs. (2, 4). Several advantages of this transformation can be mentioned. First, the original diffusion problem consisted of coupled partial differential equations in a three and a two dimensional region, whereas the integral equation is only one dimensional. Second, the boundary conditions are automatically incorporated. And third, approximate (and sometimes exact, see Eq. (8)) solutions are more easily derived. A desadvantage is that the kernel **K**, although explicitly known, is in general a complicated function of the distance $\Delta$ and the parameters which determine the boundary conditions of the original partial differential equation. This problem is discussed in Sect. 3.

The complete derivation is given in Ref. (11). Rather than reproducing it here we sketch the basic mathematical procedures involved. This both saves space and gives a transparent illustration of the relatively complex methods used.

The presupposition in Sect. 1 of a diffusion boundary layer leads to the conclusion that for all real $x$ and $y$ and for $z$ between 0 and $\delta$ a concentration field $C(x, y, z)$ can be defined which satisfies Laplaces equation

$$\Delta C = 0 \tag{17}$$

where $\Delta$ is the three dimensional Laplace operator. The Green's function for this differential equation and this diffusion layer is the concentration field around a unit sink of growth units, which vanishes at the boundaries (i.e. at $z = 0$ and at $z = \delta$). It can be obtained explicitly and used in a standard way[13] to express the concentration at an arbitrary point in terms of its value on the boundaries. At the upper boundary ($z = \delta$) the boundary condition

$$C(x, y, \delta) = C_\infty \equiv C_{eq}(1 + \sigma) \tag{18}$$

expresses that, immediately above the diffusion boundary layer, the actual concentration of growth units equals the average concentration $C_\infty$ of the supersaturated mother phase. Using this boundary condition it follows that the concentration $C$ at every point inside the boundary layer can be expressed as

$$C(x, y, z) = zC_\infty/\delta + \frac{1}{2\pi}\int_0^\infty\int\int_{-\infty}^{\infty} J_0(k\Delta r)k\,\frac{\sinh(k\delta - kz)}{\sinh(k\delta)}\,C(\xi, \eta, 0)d\xi d\eta dk \tag{19}$$

where $\Delta r$ is the distance between $(x, y, 0)$ and $(\xi, \eta, 0)$.

At the lower boundary ($z = 0$) two relations between $C(x, y, 0)$ and the normal derivative $\partial C(x, y, 0)/\partial z$ are next obtained. The first one follows from differentiation of Eq. (19). The second one is the boundary condition which describes the exchange of growth units between boundary layer and surface. Here two situations are distinguished.

First, at the step strip **S** (a region along the steps whose width $a$ is small compared to $\delta$) there is direct exchange between boundary layer and step. The boundary condition which holds is essentially due to Chernov[1]:

$$\Lambda_c \frac{\partial C}{\partial z}(x, y, 0) = C(x, y, 0) - C_{eq}(s)\,, \quad (x, y, 0) \in \mathbf{S}\,. \tag{20}$$

Here $\Lambda_c$ is the characteristic length of the volume-step exchange process and $C_{eq}(s)$ is the concentration of growth units which would be in equilibrium with the step at $\mathbf{r}(s)$ (taking the local curvature and stress into account). The left hand side is proportional to the net flux of growth units as given by Fick's law whereas the two right hand side terms represent the forward and backward flux.

Second, for positions $(x, y, 0)$ between steps, Eq. (20) is modified in a manner which is due to Gilmer, Ghez and Cabrera[2]. Indeed, for such points the back flux is proportional to the surface concentration $c(x, y,)$ of growth units.

$$\Lambda \frac{\partial C}{\partial z}(x, y, 0) = C(x, y, z) - \Lambda c(x, y)/(D\tau) , \quad (x, y, 0) \notin \mathbf{S} \tag{21}$$

where $\Lambda$ is the characteristic length of the volume-surface exchange process, $D$ is the volume diffusion coefficient and $\tau$ the mean residence time of growth units on the surface.

Upon substituting the boundary conditions in the general expression for gradients in the mother phase (to be found from Eq. (19)) an integral equation for $C(x, y, 0)$ and $c(x, y)$ is obtained, in which $C_{eq}(s)$ appears as a known and $C(s) = C(x(s), y(s), 0)$ as an unknown function.

In a similar way a second equation for $C(x, y, 0)$ and $c(x, y)$ can be obtained from a consideration of the differential equation for the surface concentration:

$$\lambda^2 \nabla^2 c(x, y) = c(x, y) - D\tau C(x, y, 0)/\Lambda \tag{22}$$

where $\lambda$ is the mean displacement of adsorbed growth units during their residence time $\tau$. The Green's function for this equation and the region between the step strips would be very complicated and it would depend on the actual step configuration. However, as long as the width $a$ is small compared to the step spacings and to $\lambda$, it makes no essential difference for the surface diffusion fields if $a$ is taken equal to 0. In that case the Green's function for the infinite two dimensional plane can be used. It is[13] the well known potential function $K_0(\Delta/\lambda)/(2\pi)$. This allows us to express the surface concentration $c$ in terms of $C(z = 0)$ and the difference $[\partial c/\partial n](s)$ of the gradients in the normal directions to the step in the plane. We assume that the concentration difference on both step sides can be neglected ($[C](s) = 0$) and that the normal gradients satisfy a similar boundary condition as in the mother phase:

$$\Lambda_s[\partial c/\partial n](s) = c(s) - c_{eq}(s) , \tag{23}$$

where $c_{eq}(s)$ plays on the surface the same role as $C_{eq}(s)$ in the volume. Hence the expression for $c$ is another integral equation for $c(x, y)$ and $C(x, y, 0)$ but now $c_{eq}(s)$ appeas as a known and $c(s)$ as an unknown function.

Next a Fourier transformation on the two integral equations for $C$ and $c$ leads to two linear equations for their Fourier transforms. Solving this set and transforming back one ends up with a coupled set of integral equations for the surface and volume concentrations $c(s)$ and $C(s)$ in the immediate neighbourhood of the step. These concentrations, however, are, by the boundary conditions Eqs. (20–23) related to the normal gradients. The latter determine, by Fick's law, the flux of growth units to the step. Indeed the contributions $v_s(s)$ and $v_v(s)$, from the surface and from the volume diffusion fields respectively, to the step velocity $v(s)$ are related to $c(s)$ and $C(s)$ by

$$v_s(s) = f_0\lambda^2[\partial c/\partial n](s)/\tau = \{c(s) - c_{eq}(s)\}f_0\lambda^2/(\tau\Lambda_s) \,, \tag{24}$$

$$v_v(s) = af_0 D\partial C(s)/\partial z = \{C(s) - C_{eq}(s)\}af_0 D/\Lambda_c \,, \tag{25}$$

$$v(s) = v_s(s) + v_v(s) \,, \tag{26}$$

where $f_0$ is the area covered by a growth unit. Using Eqs. (24, 25) to express $c(s)$ and $C(s)$ in terms of $v_s(s)$ and $v_v(s)$ we finally obtain the basic integral equations for $v_s$ and $v_v$. They can be written symbolically in the form of Eq. (4), or, somewhat more explicitly, as

$$v_s + K^{ss}v_s + K^{sv}v_v = v_{ms} \,, \tag{27}$$

$$v_v + K^{vs}v_s + K^{vv}v_v = v_{mv} \,, \tag{28}$$

In order to proceed further, we have to be more specific, both about the kernel functions $K^{\alpha\beta}$ and about the maximal velocities $v_{m\alpha}$. This can best be done in terms of the dimensionless parameters given in Table 2 (or an equivalent set). The parameter $b = \lambda/\Lambda$ was introduced in Ref. (2) to describe the importance of volume surface coupling. In analogy $b_c = a/\Lambda_c$ measures the lateral volume gradients due to the direct integration process. Finally $q = \lambda/\Lambda_s$ measures the influence of step movement on surface gradients and $\phi = \Lambda_c/\Lambda$ the relative impedance for direct and indirect incorporation of growth units. Needless to say that these interpretations of the dimensionless parameters give a rough impression only.

We start to discuss the kernels $K^{\alpha\beta}$. It turns out that they are built up from three different types of integrals. We shall refer to these integrals as bare volume potential $K^1$, bare mixed potential $K^0$ and bare surface potential $K^{-1}$. The reason to use these superscripts will become clear in Sect. 3.1, where we have transformed the integrals in a series expansion.

They are defined as

$$K^{-1}(\Delta) = \int_0^\infty J_0\left(\frac{k\Delta}{\lambda}\right)\left[1 + \frac{k}{b}\coth\frac{k\delta}{\lambda}\right]g(k)dk \,, \tag{29}$$

$$K^0 = \int_0^\infty J_0\left(\frac{k\Delta}{\lambda}\right)g(k)dk \,, \tag{30}$$

$$K^1(\Delta) = \int_0^\infty J_0\left(\frac{k\tilde{\Delta}}{\lambda}\right)[1 + k^2\lambda^2]g(k)dk \,. \tag{31}$$

**Table 2.** Coupling parameters and pseudo Nusselt numbers

| | | |
|---|---|---|
| $b$ | $= \lambda/\Lambda$ | lateral volume gradients due to surface gradients |
| $b_c$ | $= a/\Lambda_c$ | lateral volume gradients due to direct incorporation |
| $q$ | $= \lambda/\Lambda_s$ | surface gradients due to step motion |
| $\phi$ | $= \Lambda_c/\Lambda$ | relative impedance of volume-step and volume-surface incorporation |
| $\tilde{N}u$ | $= \delta/\Lambda$ | relative impedance of volume diffusion and volume-surface incorporation (= $Nu$ if $a = 0$ and surface and step processes are fast) |
| $Nu$ | | Nusselt number: ratio of surface and diffusional contribution to the crystal growth rate |

where $\Delta = \Delta(s', s)$ is the distance between the step points at $\mathbf{r}(s)$ and at $\mathbf{r}(s')$, and $\tilde{\Delta}$ an effective distance given by

$$\tilde{\Delta}^2 = \Delta^2 + a^2/4 \ . \tag{32}$$

In all expressions a factor $g(k)$ appears which tends to $\delta k/\lambda$ for small $k$ and to $k^{-2}$ at large $k$:

$$g(k) = [bk + (1 + k^2) \coth (k\delta/\lambda)]^{-1} \tag{33}$$

The coupling parameter $b$ and several other ones which will be used further on are given in Table 2. Finally we express the kernels $K^{\alpha\beta}$ in terms of the bare potentials $K^{\alpha}$:

$$2\pi\lambda K^{ss} = bqK^{-1} + b_c K^0 \tag{34}$$

$$2\pi\lambda K^{sv} = b^2(1 - \phi^{-1})K^0 \tag{35}$$

$$2\pi\lambda K^{vs} = b_c K^0 + b_c^2 \phi/(bq) K^1 \tag{36}$$

$$2\pi\lambda K^{vv} = b_c(1 - \phi)K^1 \tag{37}$$

Note that if direct integration is prohibited ($b_c = 0$) then the pure surface interaction $K^{ss}$ is proportional to the bare surface potential $K^{-1}$, and if indirect integration does not occur ($b = \phi = 0$) then the pure volume potential $K^{vv}$ is proportional to the bare volume potential $K^1$.

Finally we give the form of the maximal velocity $v_m$. As mentioned in Sect. (2.1) it would be the velocity of the step in the absence of diffusional interaction. It is proportional to the local supersaturation. Both contributions $v_{mv}$ and $v_{ms}$ contain a factor:

$$v^* = f_0 D[C_\infty - C_{eq}] \ , \tag{38}$$

which can be estimated with reasonable accuracy for most experimental situations. That is not the case with the coupling factors and the critical radius of curvature $r_c$ which are, together with the actual radius of curvature $\varrho(s)$, necessary to give

$$v_{ms} = v^* bq[1 - r_c/\varrho(s)] \tag{39}$$

$$v_{mv} = v^* b_c[1 - r_c/\varrho(s)] \tag{40}$$

for a curved step in the absence of stress. Further modifications due to stress are given, e.g., in Refs. (14, 15).

# 3 Diffusional Interaction Strength

## 3.1 The Interactions $K^\alpha$ Between Points of a Step Pattern

In this section we discuss the analytical properties of the kernels in some detail. The reason to do this rather carefully is that it turns out that the convergence of the integrals defining them is relatively slow such that the limiting behavior in different situations is sometimes difficult to subtract from the definitions.

The first point which is observed is that in Eqs. (29–31) the surface and mixed potentials $K^{-1}$ and $K^0$ are given in terms of the real distance $\Delta$ but the volume potential $K^1$ in terms of the modified distance $\tilde{\Delta}$. Following the derivation[11], sketched in Sect. 2.5. one finds $\tilde{\Delta}$ everywhere. Since, however, the width $a$ of the step strip always has to be small we may replace $\tilde{\Delta}$ by $\Delta$ unless this leads to unphysical results. In practice the latter happens with the integral $K^1$ only. Indeed one always has to integrate the $K^\alpha(\Delta)$ over step lines as in Eq. (13), in particular along the step on which the fixed step point lies. The common technique is to interchange the integrations over $\Delta$ and $k$. The integration over $\Delta$ leads to expressions $I^\alpha$, similar to Eqs. (29–31) given above, except that $J_0$ is replaced by a function which, for large $k$, oscillates with period $ak/2\lambda$ and amplitude inversely proportional to $k$. Taking the limit $a \to 0$ leaves $I^0$ and $I^{-1}$ finite (they are absolutely convergent) but makes $I^1$ infinite. The physical reason that $I^1$ behaves differently is that it is related to potentials corresponding to the three dimensional Laplace equation, which diverge too fast at small distances to be integrated. The other two are related to the surface potentials which diverge only logarithmically at small distances and hence can be integrated over an interval of distances around $\Delta = 0$.

In order to reduce the complexity of $K^1$ it is useful to note that it can be split into two terms with some elementary algebra:

$$K^1 = \int_0^\infty J_0\left(\frac{k\tilde{\Delta}}{\lambda}\right) \tanh\left(\frac{k\delta}{\lambda}\right) dk - b\int_0^\infty J_0\left(\frac{k\Delta}{\lambda}\right) kg(k) \tanh\left(\frac{k\delta}{\lambda}\right) dk \tag{41}$$

The first term corresponds to the Chernov limit of direct integration only ($b = 0$) and in the second term $\tilde{\Delta}$ could be replaced by $\Delta$ now.

Similarly, from $K^{-1}$ the term corresponding to pure surface diffusion (the Burton, Cabrera and Frank limit, $\delta = 0$) can be split off and even integrated explicitly[16] to give

$$K^{-1} = K_0\left(\frac{\Delta}{\lambda}\right)/b + \int_0^\infty J_0\left(\frac{k\Delta}{\lambda}\right) \frac{g(k)}{1 + k^2}\, dk\ . \tag{42}$$

The form given above enables us to deduce the limiting behaviour of the kernels at short distances. Indeed if the distance $\Delta$ is small compared to the characteristic length scales $\delta$ and $\lambda$ we may replace the tanh and coth by unity and then it follows that:

$$K^1 \approx (\tilde{\Delta}/\lambda)^{-1} - b\, K_0(\Delta/\lambda) + \ldots \tag{43}$$

$$K^0 \approx \frac{\pi}{2} - \frac{\Delta}{\lambda} + \ldots \tag{44}$$

$$K^{-1} \approx K_0\left(\frac{\Delta}{\lambda}\right)/b - 1 - \frac{\Delta}{2\lambda b} + \dots \tag{45}$$

This dependence on $\Delta$ at short distances could have been expected. Indeed the kernels $K^{ss}$ and $K^{vv}$ are proportional to the concentration fields on the surface and in the volume due to a unit source of growth units. At short distances it follows from Eqs. (17, 22) that these fields are solutions of Laplace's equation in two and in three dimensional space. It is well known that these solutions depend logarithmically and inversely proportionally on the distance.

In order to obtain the long distance behaviour of these potentials we expand them in a series which resembles a Schlömilich series. This expansion uses the following theorem of complex analysis[17]. Let g be a complex function with simple poles at $k_1, k_{-1}, k_2, k_{-2}, \dots$ and residues $b_1, b_{-1}, b_2, b_{-2}, \dots$ there; let it be possible to choose circles $C_1$, $C_2$ with centre in the origin such that their radius tends to infinity and that $|g|$ is smaller than a certain number on all these circles; then $g$ can be expanded as

$$g(k) = g(0) + \sum_n [b_n (k - k_n)^{-1} + k_n^{-1}] \tag{46}$$

This theorem can be applied to the factor g appearing in the kernels. We see that $g(0) = 0$, and it can be shown that the poles are on the imaginary axis. Upon writting $k_n = i\nu_n$ we find that the real numbers $\nu_n$ are the solution of

$$b\nu_n = (1 - \nu_n^2)\operatorname{cotg}(\nu_n\delta/\lambda) \tag{47}$$

and with some effort the corresponding residues can be given in terms of $\delta/\lambda$ and the Nusselt number $\tilde{N}u$ (see Table 2) for pure indirect integration

$$b_n = \frac{\delta}{\lambda}(1 - \nu_n^2)/\{\tilde{N}u[1 + \nu_n^2(1 + \tilde{N}u)] + [(1 - \nu_n^2)\delta/\lambda]^2\} . \tag{48}$$

It follows that $\nu_{-n} = -\nu_n$ and $b_n = b_{-n}$ and using this we get

$$g(k) = 2k\sum_{n=1}^{\infty} b_n/(k^2 + \nu_n^2) . \tag{49}$$

Substituting this in the equations which define the bare potentials $K^\alpha$ and interchanging summation and integration leads to[16]

$$K^\alpha = 2\sum_{n=1}^{\infty} b_n(1 - \nu_n^2)^\alpha K_0(\nu_n\tilde{\Delta}/\lambda) . \tag{50}$$

It is the simple dependence on $\alpha$ which suggested the notation $K^\alpha$.

In order to obtain more insight in this expression we have to study Eq. (47). Upon dividing both sides by $1 - \nu_n^2$ and writing $x_n = \nu_n\delta/\lambda$ the right hand side is cotg $x_n$ and does not depend on the parameters, whereas the left hand side is a hyperbola depending on the parameters $\tilde{N}u$ and $\delta/\lambda$.

An example of both sides of this equation is given in Fig. 2. Except in the special case that $\delta/\lambda$ is an integer times $\pi$ there will be one solution of Eq. (47) between all successive

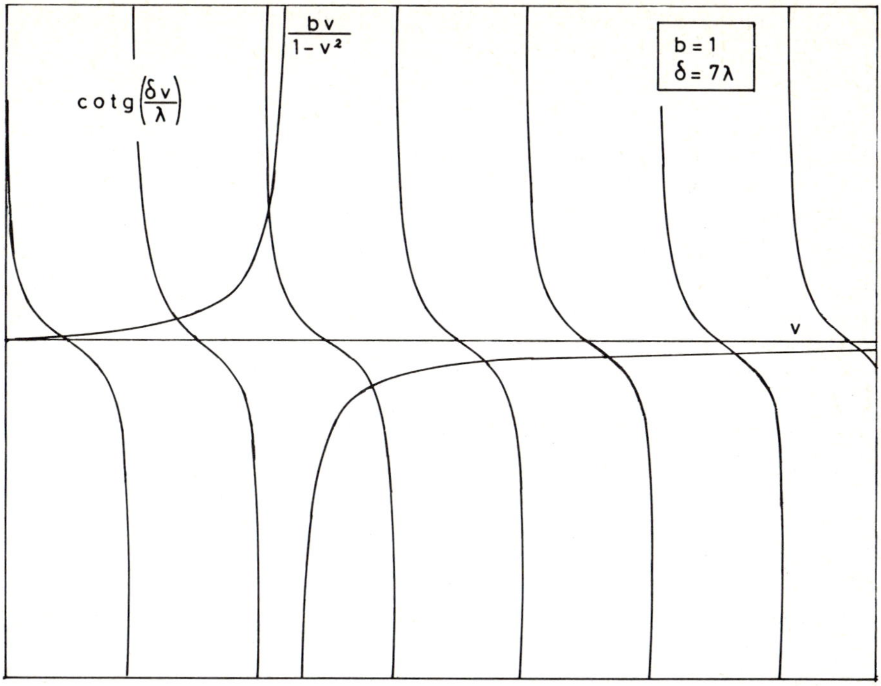


**Fig. 2.** The poles $k_n = iv_n$ of the function $g(k)$ are situated at the intersections of the hyperbola and the cotg function

zeros of the cotg and the smallest solution will be smaller than the first zero, and smaller than the asymptote at $v = 1$ of the hyperbola. In order to estimate the first pole we may either approximate cotg $x$ by $1/x$ or use the upper bound $\pi/2$ for $x_1$. Hence

$$v_1 \lesssim \min\{\pi\lambda/2\,\delta, (\tilde{N}u + 1)^{-1/2}\} \tag{51}$$

and for the other poles we have for $n = 2, 3, - \ldots$

$$\left(n - \frac{3}{2}\right)\pi\lambda/\delta < v_n < \left(n - \frac{1}{2}\right)\pi\lambda/\delta \,. \tag{52}$$

The fact that the $v_n$ are, roughly, proportional to $n$ makes the series similar to a Schlömilch expansion (in its formal definition[16]) the proportionallity should be exact and $K_0$ is replaced by $J_0$).

Note that special cases occur when $b \to 0$ or $b \to \infty$. In the first case ($b \to 0$) all poles correspond to zeros of the right hand side of Eq. (47):

$$v_1 = 1, \; v_n = \left(n - \frac{3}{2}\right)\pi\lambda/\delta \;\text{ for } n = 2, 3, \ldots \tag{53}$$

The first pole $v_1$ contributes only to $K^{-1}$ since both $b_1(1 - v_1^2)$ and $b_1$ vanish. In the second case the poles correspond to the poles of cotg $(v\delta/\lambda)$, except for the first one. We get for $b \to \infty$

$$v_1 = (\tilde{N}u + 1)^{-1/2} \tag{54}$$

$$v_n = (n-1)\pi\lambda/\delta \quad \text{for } n = 2, 3, \ldots \tag{55}$$

For large but finite $b$ values the $v_n$ again tend to the same values as for $b = 0$ when $n$ tends to infinity. These, however, do not contribute significantly since $\exp(-\tilde{d}v_n/\lambda)$ is very small.

From the integral representation of the $K^a$ it has been observed above that the narrow step limit $a \to 0$ leads to complicated results only for $K^1$ and that in that case the complications can be isolated in a more simple term. This was done by addition and subtraction of the expression for $K^1$ in case $b = 0$. Application of the same procedure to the Schlömilch expansion leads to

$$K^1 = \frac{2\lambda}{\delta}\sum_{n=1}^{\infty} K_0(v_n\tilde{\Delta}/\lambda) + \frac{2\lambda}{\delta}\sum_{n=1}^{\infty} b_n\tilde{N}u[1 + v_n^2(1 + \tilde{N}u)]K_0(v_n\Delta/\lambda)\ . \tag{56}$$

In the second term $\tilde{\Delta}$ could be replaced by the distance $\Delta$ itself, since that sum remains convergent after integrating $\Delta$ over a small interval around $\Delta = 0$ (see also the beginning of this section). The surprisingly simple term which contains the width $a$ of the step strip allows us to establish several explicit results for the pure direct incorporation mechanism. In the next section, e.g., the interaction with a single infinite straight step will be calculated exactly.

In the same way the thin layer limit $\delta = 0$ can be isolated from the bare surface potential $K^{-1}$. This gives

$$K^{-1} = \frac{2}{b}\sum_{n=1}^{\infty} \{1 - b_n[(\tilde{N}u\, v_n)^2] + [(1 - v_n^2)^2(\delta/\lambda)^2]\}K_0(v_n\Delta/\lambda)/(1 + v_n^2)\ . \tag{57}$$

From Eq. (50) it is obvious that at large distances all bare, and hence all real interaction potentials decay as Bessel functions with a characteristic length $\lambda/v_1$. Speaking loosely one might say that all potentials are cut off exponentially at a distance $\lambda/v_1$. This distance will be roughly the maximum of $2\delta/\pi$ and $\lambda(1 + \delta/\Lambda)^{1/2}$. The prefactors, however, depend both on $\alpha$ and on the system parameters. If, e.g. the boundary layer thickness $\delta \to 0$ all $v_n \to \infty$ except the smallest one, $v_1$, which tends to unity. Then the volume and mixed potentials $K^1$ and $K^0$ vanish and the surface potential $K^{-1}$ approaches $K_0(\Delta/\lambda)$, the pure Burton, Cabrera and Frank[4)] potential[18, 19)].

Finally it should be mentioned that the "Schlömilch" expansion is not useful to estimate the potentials at small distances unless it can be exactly summed. The suggestion in Ref. (2), where a similar sum has been replaced by an integral (which could be evaluated), can not be justified. Although, at small distances, $K_0$ is a slowly varying function of $n$, the prefactor changes too abruptly for such a procedure.

## 3.2 The Interaction $I_1^\alpha$ with a Straight Step

In the preceding section we have discussed how two points of a step pattern interact. In practice, however, isolated step points do not occur, only step lines can be present on crystal surfaces. It is natural, therefore, to investigate how step lines influence each other. The simplest case is the interaction with an infinite straight step, moving with uniform velocity. In view of the rapid decay of the interaction at large distances (i.e. larger than $\lambda/\nu_1$, see Sect. 3.1) this idealized case will give reasonable estimates of the step velocity with Sect. (2.4) as soon as the step is approximately straight and moves with approximately uniform velocity inside a circle of radius $\lambda/\nu_1$ around the point with which the interaction is to be evaluated.

If the step is at a distance $d$ then the distance $\Delta$ to a point of that step is given by

$$\Delta^2 = d^2 + x^2 \tag{58}$$

and the interaction with the whole step is obtained when the kernels $K^{\alpha\beta}$, or the bare potentials $K^\alpha$, are integrated from $x = -\infty$ to $x = \infty$. Denoting the integral of $K^\alpha$ by $I_1^\alpha$ we get the integral representations

$$\frac{I_1^1}{2\lambda} = \int_0^\infty \cos\left(\frac{k\tilde{d}}{\lambda}\right) \tanh\left(\frac{k\delta}{\lambda}\right) \frac{dk}{k} - b\int_0^\infty \cos\left(\frac{kd}{\lambda}\right) g(k) \tanh\left(\frac{k\delta}{\lambda}\right) dk\,, \tag{59}$$

$$\frac{I_1^0}{2\lambda} = \int_0^\infty \cos\left(\frac{kd}{\lambda}\right) g(k)\, \frac{dk}{k}\,, \tag{60}$$

$$\frac{I_1^{-1}}{2\lambda} = \frac{\pi}{2b} \exp\left(\frac{-d}{\lambda}\right) + \int_0^\infty \cos\left(\frac{kd}{\lambda}\right) \frac{g(k)}{1+k^2}\, \frac{dk}{k}\,, \tag{61}$$

and the Schlömilch expansions

$$\frac{I_1^1}{2\lambda} = \frac{\pi\lambda}{2\delta} \sum_{n=1}^\infty \nu_n^{-1} \exp(-\nu_n\tilde{d}/\lambda) - \frac{\pi\lambda}{2\delta} \sum_{n=1}^\infty \frac{b_n}{\nu_n} \tilde{N}u[1 + \nu_n^2(1+\tilde{N}u)]\exp(-\nu_n d/\lambda)\,, \tag{62}$$

$$\frac{I_1^\alpha}{2\lambda} = \pi \sum_{n=1}^\infty \frac{b_n}{\nu_n} (1-\nu_n^2)^\alpha \exp(-\nu_n d/\lambda);\ \alpha = 0, -1\,, \tag{63}$$

where $\tilde{d}$ bears the same relation to $d$ as $\tilde{\Delta}$ to $\Delta$.

The integral representations once again show why it was essential not to replace $a$ by 0 in the representation of $K^1$: at $d = 0$ the corresponding representations of $I_1^1$ would have diverged at large $k$. It turns out that the first integral in Eq. (59) can be given explicitly. That expression is found from an exact summation of the "Schlömilch" expansion in the case $b = 0$. Then the $\nu_n$ are given by Eq. (53) and the prefactor of the exponential is $\lambda/\delta\nu_n$. Hence the complete sum is the integral (with respect to $\tilde{d}/\delta$) of a geometric series. Upon summing this series and integrating the result we get finally:

$$\frac{I_1^1}{2\pi} = -\ln\tanh\frac{\pi\tilde{d}}{4\delta} - b\int_0^\infty \cos\left(\frac{kd}{\lambda}\right) g(k)\tanh\left(\frac{k\delta}{\lambda}\right) dk\ . \tag{64}$$

For the self interaction of a step ($d = 0$, $\tilde{d} = a/2$) the second term can be neglected since $a \ll \delta$. Hence, for small distances, the interaction $I_1^1$ through the volume with a straight step increases logarithmically with $d$ to $\ln(\pi a/8\delta)$, a number which tends to infinity as the step width a tends to zero. The mixed and surface interactions $I_1^0$ and $I_1^{-1}$, on the other hand, increase to a number which does not depend on $a$, but is given by Eqs. (60, 61) in case $d = 0$ (these integrals are absolutely convergent). It should be noted that the bare volume interaction $I_1^1$ diverges in the limit $a \to 0$. Nevertheless the real interactions through the volume, found from integrating $K^{vs}$ and $K^{vv}$ along the step vanish in the same limit.

For large distances to a straight step the bare interactions are, as in the preceding section, given by the first term of the Schlömilch expansion and decay exponentially, with the same characteristic length $\lambda/v_1$ as the potentials $K^{\alpha}$.

From the Schlömilch expansion the following mathematical relation between the $I_1^{\alpha}$ can be derived:

$$I_1^{\alpha+1} = I_1^{\alpha} - \lambda^2\partial^2 I_1^{\alpha}/\partial\tilde{d}^2 \tag{65}$$

This relation could be of interest when the potentials are evaluated by means of numerical methods. It shows that it is sufficient to compute $I_1^{-1}$ in dependence on $d$ since $I_1^0$ and $I_1^1$ are easily obtained. We shall not explore this possibility further.

## 3.3 The Interactions $I_\infty^{\alpha}$ with Infinite Equidistant Step Trains

In this section we write the interaction of a point on a step with an infinite equidistant step train. In this case the expressions for the interaction with a single straight step at a distance $d$ have to be summed. Taking distances $d_m = md$ to the $m$ − th step of the step strain and letting $m$ run over all positive and negative integers we get the desired interaction. Note that such a procedure together with Sect. (2.4) gives us the velocity of the whole step sequence since all step positions have the same velocity, due to the symmetry of the problem. Let us denote the resulting sums as $I_\infty^{\alpha\beta}$ and $I_\infty^{\alpha}$. We shall show how two types of series expansions for these sums can be obtained. The first one starts from the integral definitions, the other from the Schlömilch expansions of the interactions $I_1^1$, $I_1^0$ and $I_1^{-1}$ with a straight step.

For the transformation of the integral definition we use that for sufficiently regular functions $f$

$$\int_0^\infty f(k) \sum_{m=-\infty}^{\infty} \cos(kmd/\lambda)dk = \pi\lambda/d \sum_{m=-\infty}^{\infty} f(2\pi m\lambda/d) \tag{66}$$

to sum the terms where $\cos(kd/\lambda)$ appears in the integrands of the $I^{\alpha}$. For the first term in the definition of $I_1^1$, Eq. (64) we may use a definition of the function $\theta_4$:

$$\ln[\theta_4(0,\ \exp\ -\ 2x)] = \sum_{m=1}^{\infty} \ln(\tanh mx)\ . \tag{67}$$

We then get for the bare interactions

$$\frac{I^1_\infty}{2\lambda} = -\ln\frac{\pi a}{8\delta} - 2\ln\left[\theta_4\left(0,\ \exp\left(-\frac{\pi d}{2\delta}\right)\right)\right] - \frac{\pi b\lambda}{d}\sum_{m=-\infty}^{\infty} g\left(\frac{2\pi m\lambda}{d}\right)\tanh\left(\frac{2\pi m\delta}{d}\right) \tag{68}$$

$$\frac{I^0_\infty}{2\lambda} = \sum_{m=-\infty}^{\infty}\frac{1}{2m}\,g\left(\frac{2\pi m\lambda}{d}\right) \tag{69}$$

$$\frac{I^{-1}_\infty}{2\lambda} = \frac{\pi}{2b}\coth\left(\frac{d}{2\lambda}\right) + \sum_{m=-\infty}^{\infty}\frac{1}{2m}\,g\left(\frac{2\pi m\lambda}{d}\right)\Big/\left[1+\left(\frac{2\pi m\lambda}{d}\right)^2\right] \tag{70}$$

In the last expression it has been used that

$$\sum_{m=-\infty}^{\infty}\exp(-\alpha|m|d) = \coth(\alpha d/2) \tag{71}$$

which expression can be used in the Schlömilch expansions as well. Singling out the $m = 0$ term we obtain

$$\frac{I^1_\infty}{2\lambda} = -\ln\frac{\pi a}{8\delta} + \frac{\pi\lambda}{2\delta}\sum_{n=1}^{\infty} v_n^{-1}\left\{\frac{2}{\exp(dv_n/\lambda)-1} - b_n\tilde{N}u[1+v_n^2(1+\tilde{N}u)]\coth\left(\frac{dv_n}{2\lambda}\right)\right\} \tag{72}$$

$$\frac{I^\alpha_\infty}{2\lambda} = \pi\sum_{n=1}^{\infty}\frac{b_n}{v_n}(1-v_n)^\alpha\coth\left(\frac{dv_n}{2\lambda}\right);\quad \alpha = 0, -1 \tag{73}$$

These expressions can be used to obtain growth rates of misoriented surfaces. As stated in the introduction such experiments seem to be most powerful to estimate the model parameters. In the sequel we give three useful approximations.

### *3.3.1 Relatively Narrow Spacing ($\delta \gg d$)*

The argument $2\pi m\delta/d$ of the tanh and coth functions is, for $m \neq 0$, always so large that these functions can be replaced by 1 or $-1$. The only term in which the boundary layer appears is the $m = 0$ term. We separate the $m = 0$ term in each expression from the total sum and write the remainder, which depends on the parameters $b$ and $d/\lambda$ only, as $R^\alpha_\infty$. This leads to the following expressions, valid if $\delta \gg d$:

$$\frac{I^1_\infty}{2\lambda} = \frac{\pi\delta}{d} - \ln\frac{a}{d} - R^1_\infty\,, \tag{74}$$

$$\frac{I^0_\infty}{2\lambda} = \frac{\pi\delta}{d} + R^0_\infty\,, \tag{75}$$

$$\frac{I^{-1}_\infty}{2\lambda} = \frac{\pi}{2b}\coth\frac{d}{2\lambda} + \frac{\pi\delta}{d} + R^{-1}_\infty\,. \tag{76}$$

Upper bonds for the remainders can be found easily since the series defining them are monotonically decreasing and each of the terms is smaller than a constant multiple of the inverse of $1 + (2\pi m\lambda/d)^2$. One finds

$$R^1_\infty < \frac{\pi b\lambda}{d}\left(\frac{d}{2\lambda}\coth\frac{d}{2\lambda} - 1\right) < \pi b/2 \ , \tag{77}$$

$$R^{-1}_\infty < R^0_\infty < \frac{d}{2\lambda}\coth\frac{d}{2\lambda} - 1 < \pi d/2\lambda \ . \tag{78}$$

This shows that the remainders can be neglected if the step distance $d$ is smaller than a certain critical value. For the volume interaction $I^{-1}_\infty$ this critical spacing is $2\delta/b$, for the mixed interaction $I^0_\infty$ it is $2(\delta\lambda)^{1/2}$ and for the surface interaction $I^{-1}_\infty$ the maximum of $2\lambda/b$ and $2(\delta\lambda)^{1/2}$.

### 3.3.2 *Wide Spacing*

The next case to consider is the limit of widely spaced steps. To obtain approximations here it is most convenient to use the Schlömilch expansions. Since $d \gg \delta$ we have that $dv_n \gg \lambda$ for $n = 2, 3, \ldots$ Hence the second sum in the expansion of $I^1_\infty$ can be approximated by its first term and the coth in the expressions for the $I^\alpha_\infty$ by unity in the terms with $n = 2, 3, \ldots$ Whether these approximations are valid for the term $n = 1$ too depends on the ratio $\delta/\lambda$. If $\delta \gg \lambda$ then the first pole $v_1 \approx \pi\lambda/2\delta$ and hence again $dv_1 \gg \lambda$. If, on the other hand, $\delta \ll \lambda$ the first pole $v_1 \approx 1/\sqrt{(1 + \tilde{N}u)}$ and hence $dv_1 \gg \lambda$ only for distances larger than $\lambda\sqrt{(1 + \tilde{N}u)}$. From these arguments it follows that in the limit of widely spaced steps

$$I^1_\infty \approx -\ln\frac{\pi a}{8\delta} - \frac{\pi\lambda}{2\delta}\sum_{n=1}^{\infty} b_n \tilde{N}u[1 + v_n^2(1 + \tilde{N}u)] \ , \tag{79}$$

$$I^\alpha_\infty \approx \pi \sum_{n=1}^{\infty} \frac{b_n}{v_n}(1 - v_n^2)^\alpha \ ; \quad \alpha = 0, -1 \ , \tag{80}$$

and that these approximations good as long as the step spacing exceeds both $\delta$ and $\lambda\sqrt{(1 + \tilde{N}u)}$.

At this point it is of interest to note that Ghez and Gilmer[3)] replaced in Eq. (70) for $I^{-1}_\infty$ the sum by an integral. Evaluating this integral with the residue theorem leads to Eq. (80) above. They noted that the approximation is good when the poles of the function, to be summed in Eq. (70), are far enough away from the real axis. This is exactly equivalent with our demand $v_n d \gg \lambda$ for all $n$ (erroneously they stated that this would happen as soon as $d \gg \lambda$).

### 3.3.3 *Thin Boundary Layer* ($\delta = 0$)

Rather than approximating the expressions for the $I^\alpha_\infty$ in this case we use the single equation of motion, given in Sect. (4.4) for this model. Integrating $\Delta$ along the steps and summing the contributions of the steps we get for the velocity v of each step

$$v = v^* \frac{bq + b_c + 1/2\, qb_c \coth(d/2\lambda)}{1 + 1/2\, q \coth(d/2\lambda)} = v^* \left[ b_c + \frac{bq}{1 + 1/2\, q \coth(d/2\lambda)} \right] \tag{81}$$

Where the reference velocity $v^*$, given in Eq. (38) can be estimated independently with reasonable accuracy. If, therefore, the dependence of $v$ on the step spacing d is known for a sufficiently large range of d values this enables us to find the parameters $b$, $q$, $b_c$ and $\lambda$. Indeed for large spacings $v$ will be independent on d and the point where the dependence sets on estimates $\lambda$. Over the range of d values where v depends on d we may approximate $\coth(d/2\lambda)$ by $2\lambda/d$ which means that the graph of $v$ versus $2\lambda/d$ is an hyperbola from which the parameters $bq + b_c$, $qb_c$ and $q$ can be found, and consequently $b$, $q$ and $b_c$ as well. Of course in many systems it may be impossible to approach the limit $\delta \to 0$ closely enough.

### 3.4 The Interaction $I_0^\alpha$ with a Circular Step

In order to integrate the $K^{\alpha\beta}$ along a circular step one may employ Neumans addition theorem[16)] of the Bessel function $J_0$. The method is described in Ref. (11) and here we present the results for the bare potentials of a point at distance $d$ from the centre of a circular step with radius $R \neq d$. We denote the integral of $K^\alpha$ along the step by $I_0^\alpha$. The integral representations are

$$I_0^1 = 2\pi R \int_0^\infty J_0\left(\frac{kR}{\lambda}\right) J_0\left(\frac{kd}{\lambda}\right)(1 + k^2) g(k) dk\ , \tag{82}$$

$$I_0^0 = 2\pi R \int_0^\infty J_0\left(\frac{kR}{\lambda}\right) J_0\left(\frac{kd}{\lambda}\right) g(k) dk\ , \tag{83}$$

$$I_0^{-1} = 2\pi R \int_0^\infty J_0\left(\frac{kR}{\lambda}\right) J_0\left(\frac{kd}{\lambda}\right)\left(1 + \frac{k}{b} \coth \frac{k\delta}{\lambda}\right) g(k) dk\ . \tag{84}$$

Similarly one may obtain the Schlömilch expansions by application of the addition theorem for $K_0$ and one finds (for $d > R$):

$$I_0^\alpha = 4\pi R \sum_{n=1}^{\infty} b_n (1 - v_n^2)^\alpha I_0(v_n R/\lambda) K_0(v_n d/\lambda)\ . \tag{85}$$

In this expression $d$ and $R$ should be interchanged if $d < R$. If one wants to evaluate the growth rate of a circle one needs the case that the fixed point is on the circle ($d = R$) as well. Then in the expressions for $I_0^1$ one should replace $d$ by $R + a/4$ and R by $R - a/4$ in order to retain convergence.

Note that for small radii $R$ the $I_0^\alpha$ equal $2\pi R$ times $K^\alpha$. Physically this means that small circular steps behave as point sinks of growth units, with a strength proportional to their circumference.

# 4 Approximate Equation of Motion

## 4.1 Direct Incorporation Only ($b = 0$)

If $\Lambda \to \infty$ or $\lambda \to 0$ no growth units can reach the step through the surface and only direct integration of growth units from the mother phase into the step is possible (see Fig. 3). This is the model given originally by Chernov[1]. It corresponds, in our notation, to the case $b = 0$. Then, however, $v_{ms}$ and $K^{sv}$ are equal to zero and hence Eq. (27) turns into a homogenous equation for $v_s$ with only the trivial solution $v_s = 0$ (since $K(s, s') > 0$ for all $s$ and $s'$ the operator $K^{ss}$ has positive eigenvalues only).

Equation (28) is in this case a single equation which determines $v = v_v$:

$$v(s) + (2\pi)^{-1} b_c \int \int_0^\infty J_0(k\tilde{\Delta}) \tanh(k\delta) dk\, v(s') ds' = v_{mv}(s) , \qquad (86)$$

since $K^{vv}$ is proportional to the base volume potential $K^1$ given by Eq. (31), (41) or (50) ($k/\lambda$ in Eq. (41) is replaced by $k$ here).

Note that the kernel $K^{ss}$ is non-zero but nevertheless does not appear in the analysis. Physically this means that even if there is no exchange between surface and mother phase (i.e. $\Lambda \to \infty$) there is an interaction through the adsorbed growth units if their mean displacement $\lambda$ does not vanish. In a steady state situation, however, these adsorbed growth units cannot lead to step movement because there is no supply from the mother phase.

Upon assuming that $v^*$ can be estimated with acceptable accuracy, we see that the direct incorporation model depends on the parameters a, $\delta$ and $b_c$ which should be obtained by interpretation of experimental observations. Since rough estimations for $a$ ($\approx$ step height) and $\delta$ (from hydrodynamics) are available one would conclude that indirect integration is significant if the calculated and estimated values are considerably different.

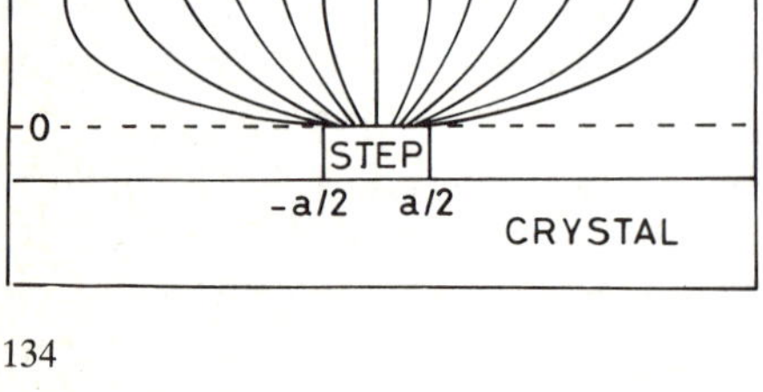

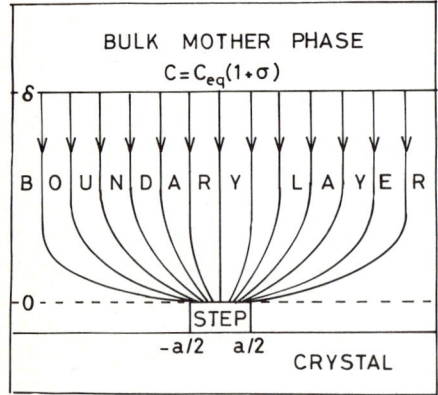

**Fig. 3.** The Chernov limit of direct incorporation ($b = 0$)

## 4.2 Indirect Incorporation Only ($a = 0$)

If the width $a$ of the step strip vanishes, the growth units reach the step through the surface only (see Fig. 4). This indirect mechanism has been studied first by Gilmer, Ghez and Cabrera[2]. Taking $b_c$ equal to zero we find that $v_{mv}$ and $K^{vs}$ vanish such that Eq. (28) turns into a homogenous equation of $v_v$ with only the trivial solution $v_v = 0$. Equation (27) is in this case the single equation which determines $v = v_s$:

$$v(s) + \frac{bq}{2\pi\lambda} \int K^{-1}(s, s')\, v(s')ds' = v_{ms}(s) \tag{87}$$

since $K^{ss}$ is a constant multiple of the bare surface potential $K^{-1}$ given by Eq. (29), (42) or (50).

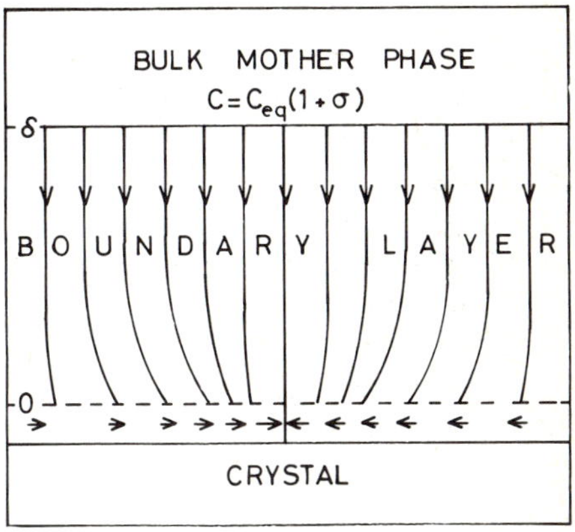


**Fig. 4.** The Gilmer, Ghez and Cabrera limit of indirect incorporation ($a = 0$)

A special case of this model occurs when not only the step strip is infinitely narrow but the boundary layer becomes infinitely thin as well (see Fig. 5). This corresponds to the Burton, Cabrera and Frank model[4] and we have[18, 19]

$$v(s) + \frac{q}{2\pi\lambda} \int K_0\left(\frac{\Delta}{\lambda}\right) v(s')ds' = v_{ms}(s) \tag{88}$$

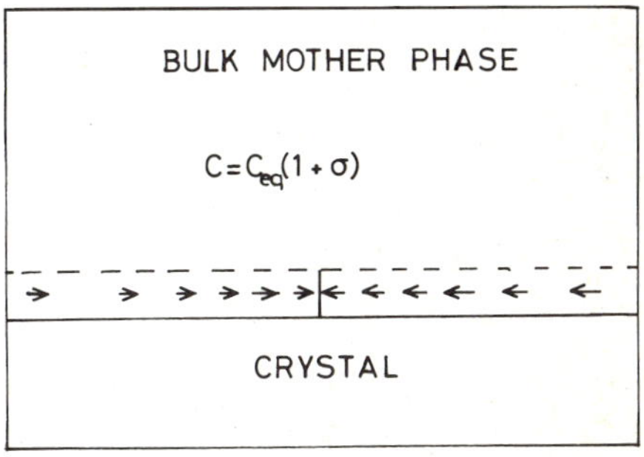


**Fig. 5.** The Burton, Cabrera and Frank limit of pure surface diffusion ($a = \delta = 0$)

Upon assuming that $v^*$ can be estimated with acceptable accuracy, we see that the indirect incorporation model depends on the parameters $b$, $q$, $\delta$ and $\lambda$ which should be obtained by interpretation of experimental observations. Since hydrodynamics provide a rough estimate of $\delta$ one would conclude that direct integration is significant if the calculated value of $\delta$ differs considerably from this estimate.

## 4.3 Equivalent Incorporation ($\phi = 1$)

When the activation energies for indirect incorporation of growth units in the surface layer and for direct incorporation in the step are equal we have $\Lambda = \Lambda_c$ or $\phi = 1$. In this case $K^{sv}$ and $K^{vv}$ vanish. Consequently Eq. (27) contains only the indirect contribution $v_s$ to the velocity and is of the form of Eq. (87) for the pure indirect mechanism. Having found $v_s$ from this equation the volume correction is given by

$$v_v(s) = v_{ms}(s) - \int K^{vs}(s, s')v_s(s')ds' \ . \tag{89}$$

Physically this approximation will be realistic if the incorporation process is insensitive for the environment where the process takes place (e.g. mono component vapour deposition, solution growth with large hydratation shells etc.). In such a case it is natural to assume that the barriers for leaving the surface into the mother phase and into the step are approximately equal as well. If $a_0$ is the single jump distance for growth units diffusing over the surface this means that $a_0\Lambda_s$ equals $\lambda^2$ or

$$b_c\phi/bq \approx a/a_0 \ . \tag{90}$$

Upon assuming that $v^*$ can be estimated with acceptable accuracy we see that the equivalent integration model depends on the parameters $b$, $b_c$, $q$, $a$ and $\delta$, which should be obtained by interpretation of experimental observations. Since rough estimations of $a$ ($\approx$step height) and $\delta$ (by hydrodynamics) are available one would conclude that direct and indirect incorporation are not equivalent if the calculated values of $a$ and $\delta$ deviate considerably from these estimates. Moreover one would be sceptical if the relation Eq. (90) above is not approximately satisfied.

## 4.4 Thin Boundary Layer ($\boldsymbol{\delta} = 0$)

If the boundary layer becomes infinitely thin (see Fig. 6) the interactions through the volume $K^1$ and $K^0$ will be blocked, and only the pure surface potential $K^{-1}$ remains finite. From its definition one sees that in this limit $\lambda g(k)$ tends to $k\delta/(1 + k^2)$. Substituting this into the definitions of the bare potentials it is found indeed that $K^1$ and $K^0$ vanish and that $K^{-1}$ approaches the Burton, Cabrera and Frank[4] potential[19]. The set of equations is then

$$v_v(s) = v_{mv}(s) \tag{91 a}$$

$$v_s(s) + \frac{q}{2\pi\lambda} \int K_0\left(\frac{\Delta}{\lambda}\right) v_s(s')ds' = v_{ms}(s) \tag{91 b}$$

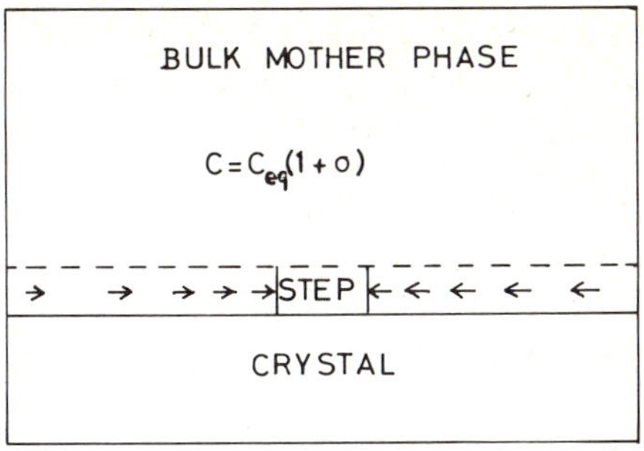


**Fig. 6.** The thin layer model ($\delta = 0$)

and they can be combined to a single equation for v

$$v(s) + \frac{q}{2\pi\lambda} \int K_0\left(\frac{\Delta}{\lambda}\right) v(s')ds' = v_m(s) + \frac{q}{2\pi\lambda} \int K_0\left(\frac{\Delta}{\lambda}\right) v_{mv}(s')ds' \tag{92}$$

This equation of motion has the same form as the Burton, Cabrera and Frank limit[18, 19]. The maximal velocity, however, is a more complicated function of the step pattern.

This thin boundary layer model is useful when experimental data suggest that they can be extrapolated to the case $\delta = 0$. Upon assuming that $v^*$ can be estimated independently with sufficient accuracy we see that the thin boundary layer model depends on the parameters $b$, $q$, $b_c$ and $\lambda$, which can, in principle, be found from growth rate of misoriented faces (see Sect. 3.3.3). These data may justify the use of one of the models mentioned in the sections above when $\delta$ is finite.

# 5 Approximate Solutions of the Equation of Motion

## 5.1 General Step Pattern

In the theory of linear integral equations several methods for approximate solutions are known. The first, most obvious, one is the Neumann series of successive iterations. In order to explain this method write Eq. (4) in the form which suggests the following iteration

$$\mathbf{v}_{n+1} = \mathbf{v}_m - \mathbf{K}\mathbf{v}_n , \tag{93}$$

where $\mathbf{v}_n$ is the iterated solution after $n$ iteration steps. Taking the maximal velocity $\mathbf{v}_m$ as the first step we get the final solution

$$\mathbf{v} = \sum_{n=0}^{\infty} (-\mathbf{K})^n \mathbf{v}_m = \mathbf{v}_m - \mathbf{K}\mathbf{v}_m + \ldots \tag{94}$$

This series converges if the norm[13] of the kernel $\mathbf{K}$ is less than unity. This norm depends both on the model parameters and on the step pattern. Roughly speaking it will be small if the coupling parameters appearing in the definitions of the $K^{\alpha\beta}$ are small. Consequently we may expect a useful approximation from a finite number of iterations only in the case of small coupling parameters. In that case the repeated operators $\mathbf{K}^2$, $\mathbf{K}^3$, ... are of higher order in these small parameters. Consequently this approximation should be used only if the diffusion interaction is small, hence the actual step velocity is close to the maximal one $\mathbf{v}_m$, and then a few steps already give a good approximation. It should be mentioned that, if this iteration converges, then the solution can also be found by a Monte Carlo method[20].

An alternative method seems not to be described in the classical literature. It is inspired by the results in Sect. (2.4) on step patterns with a uniform velocity. Physically we may expect that in many cases step velocities will not change rapidly along the steps, and hence that the diffusional interaction with the step pattern can be reasonably approximated by the interaction with a pattern which is geometrically the same but moves with uniform velocity. As before we denote with $\mathbf{I}(s)$ the integral of $\mathbf{K}(s, s')$ over all points $\mathbf{r}(s')$ of the total step pattern (see Eq. (13)). Approximating $(\mathbf{K}\mathbf{v})(s)$ by $\mathbf{I}(s)\mathbf{v}(s)$ and writing the correction term at the right hand side of Eq. (4) we are led to consider the following iteration:

$$(\mathbf{E} + \mathbf{I})\mathbf{v}_{n+1} = \mathbf{v}_m + (\mathbf{I} - \mathbf{K})\mathbf{v}_n \tag{95}$$

where $\mathbf{E}$ is the two dimensional unit operator. Since, for given $s$, both $\mathbf{E}$ and $\mathbf{I}$ do not depend on $s'$ it is easy to construct the inverse operator $(\mathbf{E} + \mathbf{I})^{-1}$. The solution can be written as

$$\mathbf{v} = \sum_{n=0}^{\infty} [(\mathbf{E} + \mathbf{I})^{-1}(\mathbf{I} - \mathbf{K})]^n (\mathbf{E} + \mathbf{I})^{-1} \mathbf{v}_m = (\mathbf{E} + \mathbf{I})^{-1} \mathbf{v}_m + \ldots \tag{96}$$

Contrary to the Neumann iteration, the convergence of this series is not determined in the first place by the norm of the operator $(\mathbf{E} + \mathbf{I})^{-1}(\mathbf{I} - \mathbf{K})$ but by the variation of the maximal velocity along the given step pattern. If, e.g., $\mathbf{v}_m$ is constant, then already the first, $n = 0$, term gives the exact solution, whatever large the operator may be. This, indeed, is the solution given in Sect. (2.4). It seems reasonable that in the physical case, where $\mathbf{v}_m$ usually varies slowly along the steps, the first term gives a reasonable approximation.

A third method to solve the integral equations approximately can be of numerical interest[21]. In that case the integrals are approximated by finite sums and the integral equations by sets of linear equations.

Finally, for sufficiently simple step patterns, the methods which expand the solution $\mathbf{v}$ in terms of eigenfunctions of $\mathbf{K}$, or in another complete set of functions can be useful. We shall not elaborate on such methods here since they are discussed in enough detail in the literature[13, 21].

## 5.2 Nucleation Between Straight Steps

In order to illustrate the power of the iteration procedure given above we may consider the problem of competition between nucleation and misorientation steps, see Fig. 7, (a

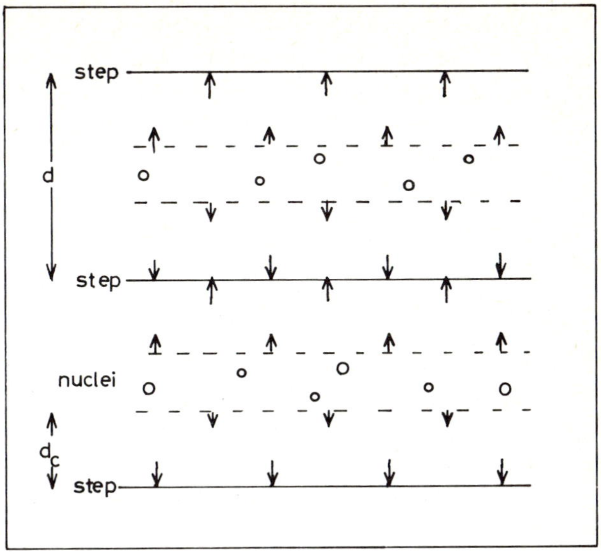


**Fig. 7.** Nuclei appear between steps when they are further than their interaction length $d_c$ apart

related adsorption experiment is described in Ref. (22)). Indeed, if the step spacing $d$ on a misoriented (vicinal) face exceeds the critical value $d_c$ where the step interaction is negligeable, then the crystal will start to generate new steps by nucleation in the middle between the steps. At the onset of this process we have, therefore, small circular steps, distributed with a line density $\varrho$ along lines in the middle between the step lines of the misoriented face. As mentioned in Sect. (3.4) each of these nuclei behaves approximately as a point sink of strength $2\pi R$. The interaction with one row of nuclei is then approximately $2\pi R$ times the interaction $I_1^\alpha$ with a straight step.

For a point on a step the function $I^\alpha$ is given by

$$I^\alpha(\text{step}) \approx \sum_{n=-\infty}^{\infty} I_1^\alpha(nd) + 2\pi\varrho R \sum_{n=-\infty}^{\infty} I_1^\alpha[(n+1/2)d] \tag{97}$$

and for a nucleus between the steps by

$$I^\alpha(\text{nucleus}) \approx 2\pi\varrho R \sum_{n=-\infty}^{\infty} I_1^\alpha(nd) + \sum_{n=-\infty}^{\infty} I_1[(n+1/2)d] \tag{98}$$

The first sums haved been evaluated and written as $I_\infty^\alpha$ in Sect. 3.3. The second sum can be found similarly when it is used that

$$\sum_{m=-\infty}^{\infty} \int_0^{\infty} f(k)\cos(k(m+1/2)d/\lambda)dk = \pi\lambda/d \sum_{m=-\infty}^{\infty} f(2\pi m\lambda/d)\cos(\pi m\lambda/d)\ , \tag{99}$$

$$\sum_{m=-\infty}^{\infty} \ln\tanh((m+1/2)x) = \ln[\theta_4(0,\exp-x)] - \ln[\theta_3(0,\exp-x)]\ , \tag{100}$$

$$\sum_{m=-\infty}^{\infty} \exp(-\alpha(m+1/2)d) = 1/\sinh(\alpha d/2)\ . \tag{101}$$

Using the same assumptions as in Sect. 3.3.1. we get for $\delta \gg d$

$$\sum_{n=-\infty}^{\infty} I_1^1((n + 1/2)d) \approx 2\pi\delta\lambda/d + \lambda \ln(\pi d/16\,\delta) \tag{102}$$

$$\sum_{n=-\infty}^{\infty} I_1^0((n + 1/2)d) \approx 2\pi\delta\lambda/d \tag{103}$$

$$\sum_{n=-\infty}^{\infty} I_1^{-1}((n + 1/2)d) \approx \frac{\pi\lambda}{b\sinh(d/2\lambda)} + 2\pi\delta\lambda/d \tag{104}$$

Upon comparing these interactions with the $I_\infty^a$ for the steps alone it can be observed that the bare volume interaction is a usually insignificant amount $\lambda \ln\delta/d$ smaller, the bare mixed interaction is the same and the bare surface interaction is $\pi\lambda/b$ smaller at step spacing exceeding $2\lambda$. Ignoring these differences we see from these considerations that the interaction matrix **I** both for the straight steps and for the row of nuclei in between is approximately a factor $1 + 2\pi\varrho R$ larger than the original matrix $\mathbf{I}_\infty$ for the steps alone:

$$\mathbf{I} \approx (1 + 2\pi\varrho R)\mathbf{I}_\infty \tag{105}$$

Let us discuss the phyiscal consequences of these mathematical results. First let us assume that we are dealing with interacting steps. Then we know that the interactions **I** dominate the unit matrix **E**. In that case the first term of Eq. (96) predicts that the step velocity is reduced with a factor $1 + 2\pi\varrho R$. The corresponding decrease in the growth rate of the crystal face is only partly balanced by the additional steps (leading to an increase of the total step length with the same factor $1 + 2\pi\varrho R$), since the new curved steps move a factor $1 - r_c/R$ slower than the original straight steps. But this is precisely what one intuitively expects: as long as steps have significant diffusional interaction they absorb all available growth units and production of new steps will not lead to an increase of the growth rate. If, on the other hand, the steps are not interacting any more then the interaction matrix **I** is small compared to the unit matrix **E** and the velocity will be reduced by a factor which is considerably less than $1 + 2\pi\varrho R$. Since still the total step length is increased by that factor it is now favourable to add the new steps.

## 5.3 The Growth Spiral

It is illustrated in Fig. 8 that the relation

$$v(s)ds = \omega r dr \tag{106}$$

should hold for a step pattern which rotates with a fixed shape around the origin. Here $\omega$ is the rotational frequency and $r = r(s)$ the distance to the centre. Upon substituting this relation in Eq. (4) we get for a spiral shape which can be parametrized as $s(r)$

$$\omega r_0 dr_0/ds + \omega \int_0^\infty K(s(r_0), s(r))r dr = v_m(s(r_0)) \,, \tag{107}$$

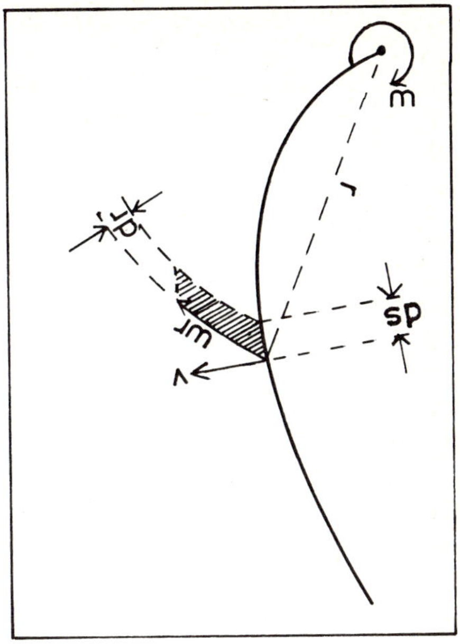

**Fig. 8.** The relation $vds = \omega r dr$ for a rigid spiral, rotating with frequency $\omega$

when we are in the limiting cases of Sects. 4.1, 4.2 or 4.4 where the problem is one dimensional. This is an equation for the shape $s(r)$. It is supposed that only for one value of $\omega$ the solution is such that the corresponding spiral shape is physically acceptable (i.e. no loops, turning points or intersections). The most relevant data which one would like to obtain from the solution are the physical rotation frequency $\omega$ (since it is proportional to the growth rate) and the step spacing $d$ far from the centre (which is proportional to the slope of the growth hillock). Approximate values of these quantities can be obtained by approximating the actual spiral by an archimedean one.

In that case we have a relation between the spacing $d$ of steps far away from the centre and the central radius of curvature $\varrho_0$:

$$d = 4\pi\varrho_0 \tag{108}$$

A second relation expresses that far away from the centre the steps should move over a distance $d$ during the time $2\pi/\omega$ of one rotation

$$d = 2\pi v(d)/\omega \tag{109}$$

where $v(d)$ is the advance velocity of a step in an equidistant infinite step train of spacing $d$ which has been obtained in Sects. 2.4 and 3.3 above. The third relation is obtained by evaluation of Eq. (107) in the centre ($r_0 = 0$), which, according to Eqs. (39, 40), then reads:

$$\omega \int_0^\infty K(s(r), 0) r dr \sim v^*(1 - r_c/\varrho_0) \tag{110}$$

At this point we have to be more specific about the kernels to be used. We have reduced the two coupled integral equations to a single one only in the limits of pure direct incorporation, of pure indirect incorporation and of infinitely thin boundary layers. In the general case the relation Eq. (106) between the rotation frequency $\omega$ and the velocity $v$ can not be used since the contributions $v_s$ and $v_v$ will, in general, not satisfy that equation. In order to obtain a first approximation, however, we will assume that close to the centre the ratio $v_s/v_v$ is a constant which is not necessarily the same as far from the centre. We expect that this approximate solution is the better, the closer the ratio $v_s/v_v$ obtained in this way for the centre is to the ratio known far from the centre. When we split the rotation frequency $\omega$ in parts $\omega_s$ and $\omega_v$ due to indirect and direct integration they satisfy

$$\omega_v \int_0^\infty K^{sv}(0, r) r dr + \omega_s \int_0^\infty K^{ss}(0, r) r dr = v^* bq(1 - r_c/\varrho_0) \tag{111}$$

$$\omega_s \int_0^\infty K^{vs}(0, r) r dr + \omega_v \int_0^\infty K^{vv}(0, r) r dr = v^* b_c(1 - r_c/\varrho_0) \tag{112}$$

The integrals are easily evaluated after interchanging the integrations over $k$ and $r$ since the latter lead to a delta function in $k$-space[11]. Upon denoting the integrals by $J_s^{\alpha\beta}$ and the corresponding bare potentials by $J_s^\alpha$ we find

$$J_s^{-1} = \delta\lambda + \lambda^2/b \tag{113}$$

$$J_s^0 = \delta\lambda \tag{114}$$

$$J_s^1 = \delta\lambda \tag{115}$$

and the values of $\omega_s$ and $\omega_v$ are given by

$$\omega_v = v^*(1 - r_c/\varrho_0)(b_c J_s^{ss} - bq J_s^{vs})/(J_s^{vv} J_s^{ss} - J_s^{vs} J_s^{sv}) \tag{116}$$

$$\omega_s = v^*(1 - r_c/\varrho_0)(bq J_s^{vv} - b_c J_s^{sv})/(J_s^{vv} J_s^{ss} - J_s^{vs} J_s^{sv}) \tag{117}$$

It is important to note that both $\omega_v$ and $\omega_s$, and hence $\omega$ as well, are, apparently, proportional to the curvature factor $1 - r_c/\varrho_0$, and that the proportionality factor is independent of the precise spiral shape. Upon substituting the expression for $\omega$ in Eq. (109) and eliminating $\varrho_0$ from this expression by means of Eq. (108) we obtain an equation for the step spacing $d$. From this equation one obtains in principle the step spacing $d$, and hence $\omega$ as well, in dependence on the model parameters[11].

As mentioned above the most "dangerous" assumption in this derivation is that the relative contributions of direct and indirect integration are constant along the whole spiral. The factor $\omega_v/\omega_s$ as obtained from Eqs. (116, 117) above can be considered as an estimate close to the centre. Far from the centre, on the other hand we have $I_s^{\alpha\beta}(\infty) = I_\infty^{\alpha\beta}$ and hence the velocities $v_v$ and $v_s$ are found from similar equations, where $J_s^{\alpha\alpha}$ is replaced by $1 + I_\infty^{\alpha\alpha}$ and $J_s^{\alpha\beta}$ by $I_\infty^{\alpha\beta}$ if $\alpha \neq \beta$. From the expressions for $I_\infty^{\alpha\beta}$ and $J_s^{\alpha\beta}$ it is seen that, as long as $\delta$ is large compared to $\Lambda$, $d$ and $bd$, $I_\infty^{\alpha\beta}$ equals, approximately, $2\pi/d$ times $J_s^{\alpha\beta}$. Consequently the ratios $\omega_v/\omega_s$ and $v_v/v_s$ are approximately equal if the $I_\infty^{\alpha\alpha}$ are large compared to unity. Using Eqs. (34–37, 74–76) it is seen that this is the case for $I_\infty^{ss}$ if $\delta$ is large compared to $(bq + b_c\phi)d$ and for $I_\infty^{vv}$ if $\delta$ is large compared $b_c(1 - \phi)d$.

In conclusion we may state that the spiral growth rate can be quantitatively estimated if $\delta$ is large and also if one of the two incorporation mechanisms is dominant. Combination of Eqs. (108, 109, 116, 117) gives estimates of the rotation frequency $\omega$ and the slope of the spiral hillocks.

## 6 Concluding Remarks

It has been the object of this chapter to illustrate the equation of motion for steps on a crystal surface, as far as this motion is determined by coupled surface and volume diffusion. It turned out that this leads to two coupled one dimensional linear integral equations. The kernels in this equation describe effective diffusional interactions.

Since this method to treat diffusion problems is new in the theory of crystal growth there is little experience about its efficiency. Probably, the most powerful property of the method is that it is, in principle, capable of estimating step velocities for arbitrary step patterns. The treatment of the growth spiral in the Burton, Cabrera and Frank[4] approximation[23] illustrated this point for the first time.

Before the method can come to its full use several steps have to be taken. First of all, one would need estimates of the parameters presented in Tables 1 and 2. As mentioned it is likely that experiments on misoriented surfaces[8–10], interpreted with the results of Sects. 2.4 and 3.3 are suited to provide these estimates in near future. Another experimental possibility to obtain the model parameters arises with the advent in near future of the in situ observation of very low, moving steps. In such a case the "solution" of the equation of motion is obtained experimentally and can be used to estimate the model parameters. It may be of interest that probably[24] growth from a gel should be interpreted as growth from a thin ($\delta \lesssim 1\ \mu m$) aqueous solution layer.

The second step is the further investigation, and the numerical evaluation of the potentials. Both the integral representations and the Schlömilch expansions could be used to this end.

And third, the methods to obtain approximate solutions have to be developed further, when complicated step patterns have to be investigated with reasonable accuracy.

After these steps have been done several applications can be imagined. The in situ observations of low steps have been mentioned already. The coupling of spiral growth and of nucleation growth with diffusional transport can be studied. Also the numerical simulation of step movement can be generalized. Finally the influence on the morphological stability of step patterns, and possibly even of crystal forms, of diffusion of growth units can be investigated.

## 7 References

1. Chernov, A. A.: Sov. Phys. Usp. *4,* 116 (1961)
2. Gilmer, G. H. et al.: J. Crystal Gr. *8,* 79 (1971)

3. Ghez, R. et al.: J. Crystal Gr. *21,* 93 (1974)
4. Burton, W. K. et al.: Phil. Trans. Roy. Soc. London *A 243,* 299 (1951)
5. Levich, V. G.: Physico Chemical Hydrodynamics, Englewood Cliffs, Prentice Hall 1962
6. Bennema, P.: Crystal Growth, an introduction, Amsterdam, North-Holland 1973
7. van der Hoek, B. et al.: J. Crystal Gr. *56,* 108 (1982)
8. van der Hoek, B. et al.: to be published in J. Crystal Gr.
9. Görnert, P. et al.: Phys. Stat. Sol. *A 57,* 163 (1980)
10. Höche, H. et al.: J. Crystal Gr. *42,* 110 (1977)
11. van der Eerden, J. P.: J. Crystal Gr. *56,* 174 (1982)
12. Höche, H. et al.: J. Crystal Gr. *33,* 246 (1976)
13. Stakgold, I.: Boundary value problems of mathematical physics, New York – London, Mac Millan 1968
14. Cabrera, N. et al.: Phil. Mag. *1,* 450 (1956)
15. van der Hoek, B. et al.: to be published in J. Crystal Gr. *56,* 621 (1982)
16. Watson, G. N.: Theory of Bessel functions, Cambridge, Univ. press 1962[5]
17. Whittaker, E. T. et al.: A course of Modern Analysis, Cambridge, University press 1952[9]
18. van der Eerden, J. P.: J. Crystal Gr. *52,* 14 (1981)
19. van der Eerden, J. P.: J. Crystal Gr. *53,* 315 (1981)
20. Hammersley, J. M. et al.: Monte Carlo Methods, London, Methuen 1964
21. Berezin, I. S. et al.: Computing Methods, Oxford, Pergamon 1965
22. Métois, J. J. et al.: J. Crystal Gr. *47,* 357 (1979)
23. van der Eerden, J. P.: J. Crystal Gr. *53,* 305 (1981)
24. Tsukamoto, K.: private communication

# Author Index Volumes 1–9

*The volume numbers are printed in italics*

Ovsienko, D. E., and Alfintsev, G. A.: Crystal Growth from the Melt. Experimental Investigation of Kinetics and Morphology. *2,* 119–169 (1979).
Pasemann, M., see Neumann, W.: *7,* 1–46 (1982).
Pfeiffer, H., see Haubenreisser, W.: *9,* 43–73 (1983).
Ploog, K.: Molecular Beam Epitaxy of III–V Compounds. 73–162 (1980).
Randles, M. H.: Liquid Phase Epitaxial Growth of Magnetic Garnets. *1,* 71–96 (1978).
Savitsky, E. M., Burkhanov, G. S., and Kirillova, V. M.: Single Crystals of Refractory and Rare Metals, Alloys, and Compounds. *7,* 107–148 (1982).
Schönherr, E.: The Growth of Large Crystals from the Vapor Phase. *2,* 51–118 (1979).
Seidensticker, R. G.: Dendritic Web Growth of Silicon. *8,* 145–172 (1982).
Sirtl, E., see Dietl, J.: *5,* 43–107 (1981).
Sugimoto, M.: Magnetic Spinel Single Crystals by Bridgman Technique. *1,* 125–139 (1978).
Takayanagi, K., see Yagi, K.: *7,* 47–74 (1982).
Tolksdorf, W., and Welz, F.: Crystal Growth of Magnetic Garnets from High-Temperature Solutions. *1,* 1–52 (1978).
Voronkov, V. V.: Statistics of Surfaces, Steps and Two-Dimensional Nuclei: A Macroscopic Approach. *9,* 75–111 (1983).
Wagner, R.: Field-Ion Microscopy in Materials Science. *6,* 1–115 (1982).
Wald, F. V.: Crystal Growth of Silicon Ribbons for Terrestrial Solar Cells by the EFG Method. *5,* 147–198 (1981).
Welz, F., see Tolksdorf, W.: *1,* 1–52 (1978).
Yagi, K., Takayanagi, K., and Honjo, G.: In-situ UHV Electron Microsopy of Surfaces. *7,* 47–74 (1982).
Zulehner, W. and Huber, D.: Czochralski-Grown Silicon. *8,* 1–143 (1982).